Ernst Cassirer: Scientific Knowledge and the Concept of Man

ERNST CASSIRER, 1874–1945

ERNST CASSIRER:
Scientific Knowledge and the Concept of Man

Seymour W. Itzkoff

UNIVERSITY OF NOTRE DAME PRESS
NOTRE DAME   LONDON

Manufactured in the United States of America

Second edition, 1997

Other titles by Seymour Itzkoff:

Cultural Pluralism and American Education, 1969
A New Public Education, 1976
Ernst Cassirer: Philosopher of Culture, 1977
Emanuel Feuermann, Virtuoso: A Biography, 1979, 1995 (2d edition)
The Evolution of Human Intelligence
    vol. 1: *The Form of Man: The Evolutionary Origins of Human Intelligence*, 1983
    vol. 2: *Triumph of the Intelligent: The Creation of Homo sapiens sapiens*, 1985
    vol. 3: *Why Humans Vary in Intelligence*, 1987
    vol. 4: *The Making of the Civilized Mind*, 1990
How We Learn to Read, 1986
Human Intelligence and National Power: A Political Essay in Sociobiology, 1991
The Road to Equality: Evolution and Social Reality, 1992
The Decline of Intelligence in America: A Strategy for National Renewal, 1994
Children Learning to Read: A Guide for Parents and Teachers, 1996

Library of Congress Cataloging-in-Publication Data

Itzkoff, Seymour W.
    Ernst Cassirer : scientific knowledge and the concept of man /
  Seymour W. Itzkoff. — 2nd ed.
      p.   cm.
    Includes bibliographical references and index.
    ISBN 0-268-00937-6 (pbk. : alk. paper)
    1. Cassirer, Ernst, 1874–1945.   2. Knowledge, Theory of.
  3. Philosophical anthropology.   I. Title.
  B3216.C34I88   1997
  193—dc21   97-22281
                                        CIP

In Memory of My Father

Contents

Preface to the Second Edition

Twenty-six years ago, when the first edition of *Ernst Cassirer: Scientific Knowledge and the Concept of Man* was published, the literature on the philosophy of Ernst Cassirer was extremely limited. The important series of translations of Cassirer's major German-language works, undertaken by the Yale University Press, had recently appeared to join Cassirer's late writings in English, *An Essay on Man* and *The Myth of the State*. There was also the classic "Library of Living Philosophers" volume on Cassirer, edited by Paul A. Schilpp,[1] to which, unfortunately, Cassirer was unable to contribute because of his sudden death in April 1945. Finally, the series of essays by Carl Hamburg, *Symbol and Reality: Studies in the Philosophy of Ernst Cassirer*, constituted the only book-length analysis of Cassirer's philosophy before the writing of the present work.[2]

Much has changed in the intervening years. My own second volume devoted to the study of Cassirer's ideas, *Ernst Cassirer: Philosopher of Culture*, appeared in 1977.[3] David Lipton's general historical account, *Ernst Cassirer: The Dilemma of a Liberal Intellectual in Germany, 1914–33*, appeared in 1978.[4] In 1979, Donald P. Verene edited a compilation of Cassirer's lectures, class notes, and projected essays entitled *Symbol, Myth, Culture: Essays and Lectures of Ernst Cassirer, 1933–1945*.[5] In 1981, James Haden's translation of Cassirer's *Kant's Life and Thought* (originally published in 1918) was brought out by Yale University Press,[6] which also published J. M. Krois's *Cassirer: Symbolic Forms and History* in 1987.[7] Krois, earlier a student of Verene at Emory University, had joined with E. W. Orth of Trier University to compile a series of previously unpublished Cassirer essays in German, *Symbol, Technik, Sprache: Aufsaetze aus den Jahren, 1927–1933*, published in 1985.[8]

Roughly coordinated with the fiftieth anniversary of Cassirer's death were several commemorative meetings and colloquia held at Weimar and Hamburg Universities in 1995 as well as the founding of the International Ernst Cassirer Society, which is headquartered at Heidelberg University and directed by Professor Enno Rudolph of the Protestant Institute for Interdisciplinary Research at Heidelberg. Under the sponsorship of the society, Felix Meiner Verlag of Hamburg undertook the publication of Cassirer's heretofore unpublished papers, as well as "Cassirer Studies," products of colloquia and prospective individual research.

The first of these volumes of unpublished Cassirer papers, now projected to encompass twenty volumes, appeared in 1995, edited by J. M. Krois and D. P. Verene, and was translated into English by Krois and published in 1996 by Yale University Press. This publication was celebrated in conjunction with an international meeting held at Yale devoted to Cassirer's contribution to the philosophy of culture. This volume is entitled and purported to be *The Philosophy of Symbolic Forms*, vol. 4, *The Metaphysics of Symbolic Forms*. These heretofore unpublished notes of Cassirer include relatively brief preliminary drafts (1928) for a possible fourth volume to follow Cassirer's then recently completed third and previously considered final volume of his *Philosophy of Symbolic Forms*. Volume 4 contains, in addition to other papers, "The Metaphysics of Symbolic Forms" (also 1928) and an exploration of phenomenological issues, "On Basis Phenomena," dating from 1940 (during Cassirer's Swedish years, 1935–1941), also possibly planned to be incorporated into a fourth and final volume of the *Symbolic Forms* series.

However, it should be emphasized that the assumption that Cassirer would have completed a fourth volume of his *Philosophy of Symbolic Forms* and in the manner projected by the material in this volume is an assumption of the editors. Cassirer did complete and publish many additional and important book-length manuscripts after the third and final volume of *Symbolic Forms* (1929).

I have presented the progression of Cassirer studies during the quarter-century period since the publication of the first edition of *Ernst Cassirer: Scientific Knowledge and the Concept of Man* to emphasize the change in orientation that studies in Cassirer's philosophy have undergone since my own work. My emphasis, which I believed to have followed in the spirit and letter of Cassirer's evolving analysis of human knowledge and action, saw his own interest in science to be preeminent. This theme is uppermost from Cassirer's earliest studies of the philosophy of Descartes and Leibniz, as evidenced in his inaugural doctoral dissertation under Hermann Cohen at Marburg University, which was published in 1902. His final writings in English, *An Essay on Man* and *The Myth of the State* (the latter not quite completed at his death), remained firmly in this scientific, empirical modality in their treatment of historical, political, and cultural phenomena.

Cassirer's intellectual evolution over this forty-year period, however, was toward an ever more penetrating examination of the complexities of cultural behavior as distinct from his earlier absorption in the philosophical status of the physical sciences. The challenge for future research presented in Cassirer's culminating neo-Kantian vision lies in the exploration of the problematic, puzzling symbolic behavior of human beings as they strive to create a civilizational context for cultural living. To further this evolving line of thinking research must be directed toward empirical issues concerning the biological origins of symbolic thought and action. The failures of rational thought and the consequent mythological debasement of the international fabric that Cassirer observed firsthand in the land of his beloved Germanic culture, with its Kantian heritage, have only added urgency to the quest for deeper scientific and philosophical understanding.

The recent trend of writing on Cassirer, due in part to his enormous corpus of published and unpublished writings, is based on traditional Central European philosophical concerns. There is no question that Cassirer framed much of his historical research and expression in the language of the European philosophical tradition, including the Hegelian semantic that was dominant in Germany in his early years. Current emphases, such as those of Verene[9] and Krois,[10] are clearly historical in linking Cassirer to Vico and Hegel. E. W. Orth tends to a more metaphysical rooting of the symbolic forms.[11] Enno Rudolph's search seems to be for a Cassirer whose historical studies can illuminate ethical issues for evangelical Christianity.[12]

By contrast, Susanne Langer, one of the most underappreciated philosophers of the twentieth century and a disciple of Cassirer's who dedicated to him her *Feeling and Form*, embarked on a very different philosophical journey.[13] In both her important *Philosophy in a New*

Key, which was stimulated by Cassirer's symbolic forms theory and adapted to a more Anglo-American intellectual context, and her final and major enterprise, *Mind: An Essay on Human Feeling*, Langer sought for human meaning and significance in the biological/anthropological sciences.[14]

My own work, following upon the writing of both of my studies of Cassirer's philosophy and undertaken in the spirit of Cassirer's Kantianism, investigated the evolution of human intelligence in four volumes.[15] The first book, *The Form of Man: The Evolutionary Origins of Human Intelligence* (1983) is dedicated to the memory of Cassirer. The final book in this series, *The Making of the Civilized Mind* (1990), attempts to place the issue of cultural symbolism and the Western civilizational ideal within the Kant/Cassirer perspective. It adopts the "symbolic forms" model, but here framed by recent scientific research. The problem Cassirer dealt with in *The Myth of the State*, namely the variance in human cultural achievement and its philosophical degradation, is likewise placed within more recent biosocial research.[16]

Our present situation in the study of Ernst Cassirer's philosophy presents the bringing together of two approaches. In one direction lies my claim that the critical philosophy derived from the Kantian synthesis—and reflected in a variety of modern sciences—leads us toward the goal of intellectual synthesis, a uniting in theory of cultural forms of thought with the physical-biological world of knowledge. This direction of Kantian thought must eschew traditional metaphysical philosophical categories.

In the other direction there seems to be a return, through several of the historical studies that Cassirer undertook, to a more traditional early twentieth-century *geistesphilosophie* model. Cassirer was not an atypical German professor of philosophy. His ambitions, grounded in a deep commitment to seeing the seminal insights of the Kantian synthesis applied to contemporary theoretical intellectual issues, did not preclude him from pursuing a profound and unique interest in historical context. This tendency and some of his lesser published works, as well as ruminative notes and lectures, constitute the source of recent interest in an Ernst Cassirer that the reader will not find in the following pages.

To support the above interpretation, as well as to lay down the gauntlet for future dialectical engagement, I quote a phrase from a lecture Cassirer gave at Yale University in 1942, based in part on lectures given at Oxford in 1934 and entitled "Hegel's Theory of the State": "I do not think we can construct a continuous process of thought by which we are led from the premises of Kant's critique to the principles and results of the metaphysics of Hegel. Instead, on the harmony between Hegel and Kant we must, to my mind, lay stress upon the fundamental, the intrinsic and ineradicable opposition between the two systems."[17]

These contemporary questions concerning the direction and philosophical meaning of Cassirer's Kantianism will be taken up in more detail in the new "Retrospective Essay" that follows, as an introduction to this study. I hope that the interested Cassirer student and scholar will find my original study both illuminating and provocative in today's intellectual climate.

My appreciation and thanks go out to Dr. James Langford, Director of the University of Notre Dame Press, for his interest in having *Ernst Cassirer: Scientific Knowledge and the Concept of Man* once more go forward to the interested lay, scientific, and philosophical communities. The assistance provided by Smith College in the publication of the new edition, especially that by Dr. Susan Bourque, Dean for Academic Development, is gratefully acknowledged.

Notes

1. *The Philosophy of Ernst Cassirer,* ed. Paul A. Schilpp (New York: Tudor, 1944).

2. Carl Hamburg, *Symbol and Reality: Studies in the Philosophy of Ernst Cassirer* (The Hague: Martinus Nijhoff, 1956).

3. S. W. Itzkoff. *Ernst Cassirer: Philosopher of Culture* (Boston: G. K. Hall-Twayne Publishers, 1977).

4. David Lipton, *Ernst Cassirer: The Dilemma of a Liberal Intellectual in Germany, 1914–33* (Toronto: University of Toronto Press, 1978).

5. *Symbol, Myth, Culture: Essays and Lectures of Ernst Cassirer, 1933–1945,* ed. D. P. Verene (New Haven, Conn.: Yale University Press, 1979).

6. Ernst Cassirer, *Kant's Life and Thought* (1918), trans. James Haden (New Haven, Conn.: Yale University Press, 1981).

7. J. M. Krois, *Cassirer: Symbolic Forms and History* (New Haven, Conn.: Yale University Press, 1987).

8. Ernst Cassirer, *Symbol, Technik, Sprache: Aufsaetze aus den Jahren, 1927–1933*, ed. E. W. Orth and J. M. Krois (Hamburg: Felix Meiner, 1985).

9. "Metaphysical Narration, Science, and Symbolic Form," *Review of Metaphysics* 47 (1993): p. 123; and with J. M. Krois, "Introduction" to *The Philosophy of Symbolic Forms*, vol. 4 (New Haven, Conn.: Yale University Press, 1996).

10. J. M. Krois, *Cassirer: Symbolic Forms and History*. New Haven, Conn.: Yale University Press, 1987; "Verene and Cassirer: On New Beginnings." *CLIO* 23 (1994): 4, pp. 423–439; and Verene, with J. M. Krois, "Introduction" to *The Philosophy of Symbolic Forms*, vol. 4 (New Haven, Conn.: Yale University Press, 1996).

11. Ernst Wolfgang Orth, "The Modern Concept of Culture as the Sign of a Metaphysical Problem" (paper presented at the symposium "New Perspectives on Ernst Cassirer" at Yale University, New Haven, Conn., October 4, 1996); Ernst Cassirer, *Symbol, Technik, Sprache: Aufsaetze aus den Jahren, 1927–1933*, ed. E. W. Orth and J. M. Krois (Hamburg: Felix Meiner, 1985).

12. "Politische Mythen als Kulturphaenomene nach Ernst Cassirer" in *Kulturkritik nach Ernst Cassirer*, ed. Cassirer Forschung, E. Rudolph, and B. O. Kueppers, Band 1 (Hamburg: Felix Meiner, 1995), pp. 143–158.

13. Susanne Langer, *Feeling and Form* (New York: Scribner's, 1953).

14. Susanne Langer, *Philosophy in a New Key* (Cambridge, Mass.: Harvard University Press, 1942); *Mind: An Essay on Human Feeling*, 3 vols. (Baltimore, Md.: Johns Hopkins University Press, 1967–1982).

15. The series, "The Evolution of Human Intelligence," is published in four volumes: *The Form of Man: The Evolutionary Origins of Human Intelligence* (New York: Peter Lang International Publishers, 1983); *Triumph of the Intelligent: The Creation of Homo Sapiens Sapiens* (1985, reprint, New York: Peter Lang International Publishers, 1989); *Why Humans Vary in Intelligence* (1987, reprint, New York: Peter Lang International Publishers, 1989); *The Making of the Civilized Mind* (New York: Peter Lang International Publishers, 1990).

16. Ernst Cassirer, *The Myth of the State* (New Haven, Conn.: Yale University Press, 1945).

17. Ernst Cassirer, *Symbol, Myth, Culture: Essays and Lectures of Ernst Cassirer, 1933–1945*, ed. D. P. Verene (New Haven: Yale University Press, 1979), p. 109.

Introduction to the Second Edition

Retrospective Essay
Cassirer, Kantianism, Philosophical Anthropology

1. Intellectual Journey: First Phase

Cassirer very early on accepted—and never abandoned—the fundamental guiding principle of Kant's transcendental philosophy, namely that the objectivity of phenomena is not given to the senses or otherwise passively received by the mind, but is the result of the mind's imposing an a priori form on the manifold given to it. An example, indeed the prototype, of such imposition of form and objectivity on a given manifold is the mind's creation of the a priori structure of Newtonian physics by the application to sense experience of the (Kantian) categories.

 . . . Cassirer rejects the identification of physics with Newtonian physics, which he regards as constituting one physical world rather than the only possible one. He holds, moreover, that man is the maker of not only one type of world, but of worlds of different types, including apart from the world or worlds of physics, the world or worlds of language, myth, religion, and art.[1]

This broad encompassing of the Kantian vision by Cassirer was similar to Kant's own movement of thinking into areas of ethics and the problem of establishing intellectual conditions for a concept of human freedom, after and in addition to the determinism seemingly imposed by Newtonian reality and the categories of rational thought that shaped this manifold into knowledge. In the case of both philosophers, their development here was both gradual and inevitable. As with Kant, who moved toward a consideration of the world of biological phenomena as well as the aesthetic apperception of human experience, Cassirer pursued the inner formative nature of all human knowledge, inexorably into the evolving modern world of knowledge.

We might view the distinction between the task that Cassirer pursued and that of Kant as revealing part of Cassirer's own inner genie. In his intellectual biography of Kant, Cassirer describes the great

Königsberger's task as flowing from a need to achieve an intellectual architectonics, an interlocking mental mastery of the world of human experience and phenomena. As Cassirer points out, the traditional picture of Kant as a lonely aesthete is warped. Kant delighted in the manifold concretia of our world, expressed a great curiosity about current events, and was a socially welcomed guest in many of the more distinguished homes of Königsberg, a lively raconteur, and, most interestingly, a broadly knowledgeable teacher at the University.

Intellectually, for the mid-eighteenth century, there was a need to order the world of knowledge and to come to terms with the overarching Newtonian vision of the physical universe. Newton died three years after Kant's birth. In Kant's lifetime, the influence of the English theorist had pressed the necessities of his conceptual structure onto the intellectual panorama of eighteenth-century life. Even as he subsequently attempted to escape from Newtonian and Aristotelian determinism, as expressed in his first Critique, Kant was building two more logical models, which, if they evaded the letter of Newton's physics, and thus metaphysics, still reflected that omniscient presence in the analyses of phenomena and claims for knowledge.

For Cassirer, there was nothing like the omnipresent power of a Newton to challenge his thinking and method. While a devoted student of Hermann Cohen for the three years of his stay in Marburg, he had already studied with Dilthey and Simmel in Berlin, and Wundt in Leipzig. Kant's influences, on the other hand, were all second hand, in the writings of Leibniz, Newton, Wolff, Hume, and Rousseau. In addition to the more cosmopolitan environment in which Cassirer grew to maturity, the enormously cultured environment of his five cousins,[2] and unlike Kant, his independence from economic want, Cassirer's milieu was one of increasing skepticism. Helmholtz had shown long before that Euclidian geometry was only a conventional and special exemplar, Riemann and Lobatchevsky having created alternative models of so-called absolute space. In physics, action at a distance was being challenged by the work of Oersted, Faraday, Maxwell, and Hertz. Light, electrical, and magnetic phenomena were shown to exhibit wave characteristics, which eventually led to the equations of field dynamics.

Indeed, Hermann Cohen was deeply involved in an attempt to salvage Kantianism from these implicit challenges. His pursuit of the "transcendental method" as the direction of all future scientific philosophy, then elaborated in his postulation of differential calculus as a logical foundation within which he sought to secure Kantian a priorism, was developed in his *Kant's Theorie der Erfahrung* (1885). This perspective was not to be a burden passed on from the teacher to his brilliant but independent disciple, Cassirer.

If the great inner vision of Kant's intellectual task remained the logical integration of an a priori perspective on the human universe of knowledge, Cassirer's path was to be a more contextual exploration, using the historical past in all its cultural richness. This heritage had deeply affected Cassirer, especially the German cultural contribution into which he could place himself and his life. German culture was then in the vanguard of contemporary scientific progress. Cassirer's study of science in this maturing 1890s setting, the flourishing of the research of such thinkers as Hertz, Mach, and Planck in the physical sciences, made it clear that what one learned as truth was contingent on the ongoing laying out of both theory and technological applications, all in process of transforming the modern world.

Cassirer's special cast of mind can be seen in his doctoral dissertation, which was devoted to the scientific underpinnings of Descartes's philosophical thought. This 100-page work, published within *Leibniz's System in Seinem Wissenschaftlichen Grundlagen*, would form the introduction to the larger study of Leibniz's scientific and mathematical thinking and as the core dimension to the latter's philosophical writings. These were published in 1902, several years after Bertrand Russell's important book on Leibniz and in the same year as Cassirer's teacher Cohen's masterful *Logic der reinen Erkenntnis*. It is clear that Leibniz's work in calculus as well as his more conventional symbolic views of the nature of scientific theorizing seemed to Cassirer to harmonize with recent scientific and philosophical perspectives on the nature of scientific hypothesis and law.[3]

Leibniz's System was rapidly followed by a series of historical studies on the nature of scientific knowledge, starting with the Renaissance, including Cassirer's preliminary analyses of the Kantian corpus, the first three volumes of *Das Erkenntnisproblem* (1906–1909). In 1910 *Substance and Function* appeared, which marks Cassirer's first systematic work in the contemporary theory of knowledge, a study and critique of nineteenth- and early twentieth-century philosophy of science relating to the nature and status of physical theory. Here, Cassirer's classical Kantianism, without the Königsberger's commitment to Newtoninan and Aristotelian categories, shines through.[4]

> To explain nature is thus to cancel it as nature, as a manifold and changing whole. The eternally homogeneous, motionless "sphere of Parmenides" constitutes the ultimate goal to which all natural science unconsciously approaches. It is only owing to the fact that reality withstands the efforts of thought and sets up certain limits, that it cannot transcend, that reality maintains itself against the logical leveling of its content; it is only by such opposition from reality, that being itself does not disappear in the perfection of knowledge. . . . The identity, toward which thought progressively tends, is not the identity of ultimate substantial things, but the identity of functional orders and correlations. . . . The inexhaustibleness of the problem of science is no sign of its fundamental insolubility, but contains the condition and stimulus for its progressively complete solution.

Cassirer here cites the writings of the French Neo-Kantian Emile Meyerson in *Identity and Reality*.[5]

The next several years saw Cassirer involved in his teaching at Berlin and working on the forthcoming edition of all of Kant's works, to be published by his cousin Bruno Cassirer. World War I was a deeply disturbing event for all Europeans, given the relatively long interregnum of peace that had concluded the nineteenth and begun the twentieth century. It brought many scholars, including Cassirer, to a reconsideration of the historical and cultural glories of the past. Products of these years were *Freiheit und Form: Studien zur Deutschen Geistesgeschichte* (1916) and *Kants Leben und Lehre* (1918). This last work, essentially completed in 1916, was dedicated to Hermann Cohen, who died shortly before its publication. It was also the final opus of the eleven volumes of the complete works of Immanuel Kant. (Cassirer was a co-editor of the series.) And lastly, *Idee und Gestalt: Fünf Aufsaetze* was published in 1921.

It should be noted that these historical/cultural works which reexamine important intellectual and cultural figures and ideas of the German and European past are often taken up by scholars today.

They are used to bring Cassirer once more into the dynamics of mid-European metaphysical concerns, either phenomenology or even traditional Hegelian historicism, both traditions highly popular among the philosophical academics during that period. And since Cassirer saw himself as part of the mainstream philosophical community, he felt a need both to attend to these issues and discussions, and on occasion to enter into the debates, as he did at Davos in Switzerland in 1929 with Heidegger and others.

These rich and beautiful supplemental writings ought not be confused with his synthetic philosophical direction. Rather, they exhibit a duality of interest of which Cassirer was entirely self-conscious. In 1919, Cassirer was appointed professor of philosophy of the newly founded Hamburg University, a liberal product of the Weimar Republic. Cassirer would be elected Rector (1930–1932) before his sudden resignation in 1933 and decision to flee Germany and the Nazi regime.

The most important work to be completed upon his move to Hamburg was *Einstein's Theory of Relativity*, published in 1921, with an acknowledgment by Cassirer of Einstein's critical comments before publication.[6] Cassirer's task was to ascertain the significance of these revolutionary theories in the physical sciences for his by now well-established critical neo-Kantian views on scientific theory making: "the purpose of the *Critique of Pure Reason* was not to ground philosophical knowledge once and for all in a fixed dogmatic system of concepts, but to open up for it the 'continuous development of science' in which there can be only relative, not absolute, stopping points."[7]

In general, Cassirer views the new geometry, the relative structure of space and time, as completely compatible with his Kantianism. The impact of his visits to the newly established Warburg Library at Hamburg University can be seen toward the conclusion of *Einstein's Theory of Relativity*. Cassirer speaks about the different perspectives of space and time given both in Newtonian physics—the movement of bodies at speeds slower than that of light—and in Einsteinian relativistic four-dimensional physics. He warns against the mathematical and physical worldviews being interpreted in absolute terms and expresses a concern with the metaphysical psychologists' interpretations acting to reduce reality to an immediacy of space and time experiences.

However, Cassirer's immediately following thought can be puzzling: "But both views prove, in their absoluteness, rather perversions of the full import of being, i.e., of the full import of the *forms* of knowledge of the self and the world." He goes on to describe the other symbolic ways of expressing space and time aside from the psychological or the physical: the historical use of chronology, itself founded on astronomy and mathematics; painting and the laws of perspective; architecture and the laws of statics; music, whose rhythmic unity issues from very different melodic structures than even the Pythagoreans had expected.[8]

Cassirer concludes his study of Einstein's relativity by expressing another enigmatic turn of phrase: "What space and time truly *are* in the philosophical sense would be determined if we succeeded in surveying completely this wealth of nuances of intellectual meaning and in assuring ourselves of the underlying formal law under which they stand and which they obey."[9]

One can safely claim that the research that Cassirer already was undertaking and that would appear successively in three volumes as *The Philosophy of Symbolic Forms* became a lifelong quest in his search to probe the breadth of symbolic expression in culture and its dialectical movement in historical time. It is important to emphasize that it was not a task of penetrating in one swoop the inner principles of symbolic thought in all its multiple varieties. Rather, the problem would be to lay out empirically the evolving semantic represented in the development of these forms of knowing. Here we would secure for our intellectual journey the wisest researchers and the light that they communicated. The ultimate possibility for a synthetic unification of symbolic forms, as an epistemological quest, would ever be conditional, dependent on existing scientific knowledge.

2. Toward a Philosophical Anthropology

Martin Buber once stated that Immanuel Kant had created a structure of knowledge that implicitly revealed the interrelationship of humans and their minds. Kant had arrived at the threshold of a de-

scription of the creature that created the forms of knowledge, elucidated in his three critiques. On this threshold stood the opportunity
to sketch in at least the outlines of a philosophical anthropology.
Kant, unfortunately, was unable to enter this new domain of philosophy and fulfill this important need.[10]

Cassirer, as he entered the middle period of his life and thought,
engaged in a new and modern academic environment at the University of Hamburg, where, in its Warburg Library, he had available
to him a wealth of empirical scientific, ethnographic resources. He
was now able to further his philosophical explorations of the long-
delayed Kantian challenge.

The first two volumes of the *Symbolic Forms* opus were devoted
to an inquiry into the inner structure and development of human
language, mythic symbolism, and religion. There was no cohesive
body of philosophical disputation in these cultural disciplines as
there had been in the realm of physical theory, no body of argu-
mentation and interactive dialogue on which Cassirer could focus his
analytical skills.

Instead there existed a rich and diverse empirical and theoretical body of research, including ethnological and field studies by
scholars working in distant and disparate cultures. Linguistic theory,
based on much comparative analysis, had already dealt with the inner
nature of language forms and their evolution over time. The same
could be said of the rich mythic literature and comparative and systematic studies of both primitive and modern philosophical religious forms.

Cassirer's approach to the empirical data in these volumes, his use
of Kantian categories of time and space as the objectifying foci in the
symbolic reification of languages as well as mythic thinking, showed
the steady influence of his earlier Kantian and scientific work. These
modalities, space, time, and number, concentrated his analysis of linguistic and mythic thinking, helping to highlight their intellectual
movements and internal structure during historical time frames and
in the context of their very diverse geographies and ethnic traditions.

Language, myth, and religion seemed to share a trend toward
greater abstraction in the character of their symbol systems over time.
Languages, for example, when compared on the basis of cultural de-

velopment, seemed to evolve more universal conceptual modalities, gradually transforming the plethora of concrete descriptive words usually expressed within the undeveloped culture's natural and human world. Cassirer likewise surveyed comparative mythic forms of expression. These similarly tended to slough off purely physical rituals and concrete totems and taboos, leading toward ever more philosophical and ethical categories within which could be focused the basic psychological vectors of the form. What remained to underlie the fundamental psychological intentionalities that created the symbolic form in the first place were certain expressive rituals, ethical mandates of behavior, a sense of the holy, and an institutional role in social life which revealed its permanence as a dimension of symbolic meaning and, eventually, knowledge.

In the third volume, completed essentially in 1928, and published the following year, the task was more complex. This book, longer and quite varied in the topics analyzed, seemed to signal a turn in Cassirer's way of looking at the question of a proliferating palette of symbolic forms—art, myth, religion, language, history. Even the subtitle, "the phenomenology of knowledge," hinted at a more intense reexamination of the symbolic process.

Influenced by the neo-Kantian psychological studies of his fellow Marburger Paul Natorp ("Allgemeine Psychologie" 1912), as well as the phenomenological thinking of Husserl and Heidegger, Cassirer undertook to delve more deeply into the meaning of symbolic thought itself. It had been important to him to probe the diverse structures of myth, religion, and language to attempt to discover patterns and trends as well as intentionalities in these symbolic forms. Eventually, however, the philosopher would have to bend to the scientific spirit of inquiry to ask deeper questions about the basic sources for the symbolic process itself, i.e., "the full import of the *forms* of knowledge of the self and the world."[11] This import could never seduce Cassirer into metaphysical rhapsodies over hidden origins and potencies beyond the ken of the empirical mind. No psychic noumena, hidden worlds, or masked ultimate entities are to be found in his phenomenological analysis.

Topics dealing with psychological issues of perception and commonsense knowing, the significance of raw physical pathologies, such as aphasia, all pressing on the thinking of ordinary humans, are the grist of this rich and complex final volume of *The Philosophy of Symbolic Forms*. Together with Cassirer's more philosophical discussion of perceptual and intuitive modes of experience and thought, these seemingly mundane comparisons with the functioning of wounded minds opened up a new world of scientific knowledge. This research, carried out by his cousin Kurt Goldstein in the Frankfurt Neurological Institute, brought to a climax a growing awareness of the larger implications of symbolic knowing. At the core of human experience, ordinary perceptual awareness, commonsense envisionment already existed as a formative symbolic element at work in the construction of the human psyche.

Indeed, there did not exist pure or primarily given, unshaped and unstructured human experiences, as claimed by the sensationalist psychologists and the phenomenological philosophers. Likewise, suppositions of a mental reality, a truer world of Being—*Sein*—beyond the contingencies of human thought and existence, constituted traditional metaphysically inflated and empty verbalization already being decisively punctured by the Vienna School of Logical Positivism.

In 1929, shortly after the publication of the third volume of *The Philosophy of Symbolic Forms*, Cassirer was elected Rector of the University of Hamburg. Clearly, he now had added academic duties. Also, the many different philosophical strands opened up by the study entailed in this third and final volume needed additional thought and study. The notes that Cassirer directed toward a concluding fourth volume are therefore fragmentary and largely undirected in focus. One is troubled to see published as supposedly complete the as yet unshaped musings and conjectures, as well as unrelated notes, of even as great a philosopher as Cassirer.[12]

A series of historical researches was undertaken and published in the succeeding years, marked as they were by the turmoil and flight from Germany in 1933. Subsequently, Cassirer led the life of a vagabond, though distinguished scholar, first in Vienna, then in Oxford, then in Göteborg, Sweden. Cassirer departed for the United States in 1941 on the last boat to pass through the Nazi blockade of neutral ships. (One of his fellow passengers was Roman Jakobson, the noted linguist.) Despite these upheavals, however, Cassirer was able to complete two new systematic works, *Determinism and Indeterminism in Modern Physics* (1936),[13] first published in Sweden, and volume 4 of his *Das Erkenntnisproblem*.[14] Both works were a product of his all too few contented years in Sweden before the final move to America.

The first of the two books, on the status of the new quantum theories in physics, was a must for Cassirer, as controversy embroiled Einstein. Einstein had long protested against the nondeterministic probability characteristics of the new science, as well as its shifting envisionment between material and wave descriptions of phenomena in space and time. Cassirer saw no conflict here with his own symbolic view of knowledge, as quantum descriptions still performed their predictive functions and proved what Cassirer had suggested in conclusion about Einstein's relativity, namely, that other theories could be expected to provide alternate descriptions of human experience. They were in no way to be thought of as descriptions of an unchanging reality that might lie beyond the ken of our symbolic envisagement of experience.

The latter volume, volume 4 of *Das Erkenntnisproblem*, mainly devoted to biological thought, marked the completion after many years of Cassirer's investigations into the evolution of modern scientific theory. This volume, now devoted to the evolution of nineteenth-century biology, clearly a topic that must have weighed heavily on his thinking subsequent to his theory of symbolic forms, was his final systematic effort before coming to Yale University in 1941. That he allowed the manuscript to remain behind in Sweden when he left for Yale (recovered by his widow only in 1946, thence to be published in translation by Yale University Press) naturally raises the question as to whether Cassirer felt that in its unfinished state it still needed more tightening and integration of thought.

The reputation of *The Philosophy of Symbolic Forms*, then as yet untranslated from its original German, stimulated much interest in the possibility of a translation or a transcription of the theory into a succinct English-language summary. Cassirer, then at Yale and in a wholly new intellectual and cultural environment, felt it wiser to expand on the themes raised in the earlier multivolume series. He would be able to reflect on his theoretical position from a new empirical and conceptual vantage point of time and place.

An Essay on Man, written in English, was published by Yale University Press in 1944, the year of Cassirer's move to a visiting professorship at Columbia University in New York. At age seventy, he had been forced to retire from Yale. The new book naturally revealed more fully than earlier works the true direction that Cassirer would have taken in his search for deeper and ever more universal meaning in human symbolic experience and knowledge, the conventionality and instrumentalism of theory itself. The increasingly Anglo-American drift of his thinking with the impact of empirical, scientific elements seems fully apparent.

This intellectual move was epitomized in the following quote from *An Essay on Man*: "Man has, as it were, discovered a new method of adapting himself to his environment . . . the symbolic system. . . . He lives, so to speak, in a new *dimension* of reality."[15] *An Essay on Man* distills the culminating direction of his long intellectual journey. In this work, Cassirer emerged from his long phenomenological laying out of the structure and processing of symbolic thought as both knowledge and behavior, science as well as the cultural forms, in their broadest expression. He had now reached a summary position, ontological in its fundamental realization, but always amenable and subject to the dictates of science. It is a perspective that now sees the problem of human knowledge, behavior, and thought in terms of the mysterious question of "man's" place in the evolutionary biological process. In asking this question, Cassirer walked through the gateway of philosophical anthropology.

All the civilizational constructs—art, the various disciplines of science, history, political thought, myth and religion, technology— are now to be seen as unique extrusions of the evolutionary process, a process that sees Cassirer joining with the continental schools of psychology and biology in rejecting a purely reductive behaviorist interpretation of human culture. Here, again, Cassirer uses his critical Kantian heritage as a bulwark against the majoritarian orthodoxy. Indeed, humans are animals. But they are unique animals, mutations away from the instinctual intentionalities of practical adaptive behavior that we discover in all animal life. What does all this mean? Only one answer is excluded a priori, one that reifies old ontological or metaphysical essences, always inoculated in their empty verbalizations against the empirical clarity of scientific evidence.

Even while *An Essay on Man* was being received with great acclaim, Cassirer, still at the height of his powers at seventy, was deeply involved in the writing of *The Myth of the State*, which, tragically, he would not be able to see published in his lifetime. (His death came suddenly, on the Columbia University campus, in April 1945.) This final book, rich as always in its historical, philosophical survey of the literature on political philosophy, raises the groundbreaking cultural and psychological themes earlier postulated by Sigmund Freud, the deeper and darker resonances in human nature that threaten the generally Apollonian outlook seen in Cassirer's envisionment of Western civilization.

Freud, who died in 1939, did not have to view the full horror of fascism in its German and Austrian incarnations, or that other horror that lay silently behind the largely impenetrable and monolithic iron curtain, the Soviet genocides. Cassirer saw most of it without being able to be a full witness to the 1945 triumph over Axis totalitarianism.

What *The Myth of the State* did recognize was the fragility of human reason, the flaw in the seeming evolutionary advance of an intellective civilization in which the forms of symbolic expression could be held in philosophical balance. It was a flaw that went beyond the recurring conflicts of the past. Here, in this most advanced state of cognitive achievements, whole nations, whole peoples could revert to the most primitive level of language, thought, and behavior, the mythic mentality.

> Modern political myths . . . undertook to change the men, in order to be able to regulate and control their deeds. The political myths acted in the same way as a serpent that tries to paralyze its victims before attacking them. Men fell victims to them without any serious resistance. They were vanquished and subdued before they had realized what actually happened.[16]

In his concluding thoughts in *The Myth of the State*, Cassirer reflected upon a new struggle in human history, one which he, as victim, now recognized as a dissonant dimension in culture and history, an element with which his philosophical anthropology, indeed any philosophical anthropology, would now have to deal.

As long as these forces, intellectual, ethical, and artistic, are in full strength, [the world of human culture] myth is tamed and subdued. But once they begin to lose their strength chaos is come again. Mythical thought then starts to rise anew and to pervade the whole of man's cultural and social life.[17]

What then are these new questions that the philosophy of symbolic forms must confront, now in terms of a truly Kantian philosophical anthropology that is attuned to the scientific and historical issues and evidence of the contemporary world? This element of weakness, of the uncontrolled emotive and expressive counterrationality that lies within the framework of the human psyche clearly must be understood as a psychic-intellective vulnerability of humans. It calls for a deeper understanding of human nature. If *Homo* has with great effort and struggle arisen from the primordial ground of raw, spontaneous, unselfconscious mythic and linguistic emotionalism, the naive barbarism that we find in primitive social and cultural environments, then we must yet painfully account for the subsequent barbarization that has afflicted the human species even as it seemingly rises toward civilizational self-consciousness and philosophical awareness. No people surging toward its acme of wealth and power can smugly ignore the serpent of irrationality that may lie within.

3. Cassirer's Kantian Legacy

Cassirer's Kantianism contains a dual legacy. First, we have the theoretical intellectual problem of a human nature that at once emanates from a mammalian/anthropoid substrate of evolutionary processes. Yet, human symbolic behavior is not reducible to the given of animal signal responses, instinctual adaptive and survivalistic motives.

Second, we have a symbolic pantheon of richly diverse cultural achievements, clearly a product of the gradual self-conscious awareness and sophistication brought about by the inner objectifying potency of the human mind. Within this progressive trend of the human species toward both the universal thrust of thought and its ever more plural psychic/symbolic/cultural intentionalities, now came unheard of tragedies in human cultural life—social degradation, genocide. Clearly, not all of this precipitating chaos and decline can be explained merely as a by-product of the mixed historical heritage of the human race. Cassirer's focus was always on the European experience as a model and paradigm of humankind's symbolic and intellectual self-liberation. Here in the very heart of his beloved and universal Germanic culture, civilization itself had stumbled, regurgitating leaders, symbols, and events of the most primitively ignorant and barbaric kind that the world had ever experienced.

These issues are not separate in terms of the construction of a truly modern Kantianism. Their unity lies in the hope for a larger overarching and assimilating theory of human nature in history. In one part of this theory the evolutionary and structural sources for symbolism need to be researched. In the second, the nature of this horrific disjunction in human behavior, the creative and abstractive versus persistent primitivism and ethical barbarity, needs to be understood. To this day, humans seem unable to evaluate their place in history and to grasp the systematic means to fulfill the potential that is revealed in human nature by the great civilizational exemplars.

To be fair, these issues do now surge to the front of our philosophical consciousness, but not necessarily as a concrete intellective challenge. Contemporary philosophical thought appears to be incapable of meeting this challenge since these issues seem to lie beyond the interests of those who define themselves as philosophers. Human nature, in its defining bio-social characteristics, has now become a domain of the social and biological sciences. Philosophers, in daring to pursue such studies, would have to continue on the path of Kant and Cassirer. They would have to immerse themselves in the sciences in a manner shown to us by Kant's own early researches in cosmology and physics.[18]

Ernst Cassirer: Scientific Knowledge and the Concept of Man was written to clarify the first question raised above, namely, what was Cassirer's contribution and challenge to a modern philosophical anthropology? Could a Kantian view of the inner integrity of the symbolic process as a core attribute of human behavior be sustained in the context of the reductive animal behavior/signal/instinct scientific model existing a quarter of a century ago? *Ernst Cassirer* attempted to demonstrate in the context of the then contemporary evolutionary, theoretical biological and psychological tradition, some thirty years after Cassirer first set the challenge in *An Essay on Man*, that indeed there was growing evidence accounting for the sharp evolutionary deviation that would explain *Homo*'s seeming non-biological behavior. At the same time, a neo-Kantian philosophical position could be contained within the mainline theoretical and evidential structures of the contemporary scientific literature on evolutionary theory.

For example, Susanne Langer's *Feeling and Form* set forth her development of Cassirer's views on art as a direct outgrowth of human self-conscious awareness of this special psychological modality of feeling, combined with the special perceptual embodiment of the particular art object or form.[19] Langer was already embarked on a larger development of these ideas. She wanted to understand the broader contextual framework within which art could be seen as a fundamental modality of human thought and behavior. Indeed, art was a symbolic creation of humans, but, as Cassirer had stated, it held forth in the realm of "non-discursive" or nonscientific symbolism. Langer's series of three volumes, *Mind: An Essay on Human Feeling*,[20] was an implicit reply to Cassirer's challenge to go to the scientific literature to secure the origins of human symbolism. Indeed, here, too, in art, a paradigmatic *geisteswissenschaftlichen* study, the origin of the artistic symbol might be more clearly revealed in the biology of animal behavior.

At the very beginning of her work, Langer raises the question probed early on by Cassirer: What is the meaning of art as a "living form"? How is it that a seemingly cultural object is imbued with so much dynamic meaning? How can it be thought to emanate from the world of life in the first place and then to enter the world of culture? In what sense does its "expressive" significance separate it from other symbolic modalities?

It is from her developed analysis in the first volume, of art as feeling and abstraction, that she delves into the biological world to present her major conceptual claim: the "act" concept. "The dynamism of life lies in the nature of acts as such; it is incorporated in their structure and gives them their typical form. Every act . . . has an initial phase, a phase of acceleration and sometimes increasing complexity, a turning point or consummation, and a closing phase or cadence."[21]

Langer will attempt to show that the art experience, both creative and consummatory, is built out of the same organic vitality as the biological or evolutionary substrate.

> Every constituent act has its particular impulse, with its own intensity and easiest path, its possible alternatives if that path is obstructed and its own rate of progress; and every impulse when it issues in action, resolves the particular organic tension it represents, which is an accumulation (perhaps infinitesimal) of potential energy ready for transformation into some other phase. In living systems such charges tend to form integrated patterns, i.e., unitary but organized impulses, and spend themselves, under the influence of one largest, unifying impulse, in a flow of events that takes the characteristic form of a single act.[22]

Ill health blocked Susanne Langer from completing her magisterial work, and only a truncated third volume appeared. However, her work provides an example of how a developed Kantian philosophical anthropology would have to operate out of a nexus of scientific and philosophical considerations. In the end, the problem of art, to which Langer so determinedly devoted herself, had to find its sources and resolution as a symbolic form as part of the Cassirerean envisionment of the original Kantian enterprise. The problem of art lies in the natural world of science, not in linguistic and thus hyper-empirical intellectual manipulations.

As Cassirer demonstrated in his own development, and congruent with the Kantian spirit, the search for philosophical meaning has to pivot on the ever changing dimensions of human existence, as the universe itself quietly produces novelty and the hitherto unknown. So, too, issues that were exploded into reality by the coming of the twentieth-century political mythologies and cultural irrationalities had now to be more intensively studied.

The last several decades, partially under the impetus of the continued conflicts and cultural unraveling experienced in the post–World War II era, saw the coming of many covert philosophical explanations. One, from scientific biology—sociobiology—interpreted the savage competitiveness for resources, personal and national dominance, as a reification of mammalian selective dynamics extruding into the human world. Are we merely elaborate chimpanzees, bipedal, carnivorous predators? Clearly, we are animals, at times vicious animals, considering the wars, the interpersonal violence, the hatred.

A mass international technological culture seems to have swamped the traditional folk and high civilizational ethos that had produced the Renaissance, the Enlightenment, and then the scientific revolutions. What is the nature of the conceit that spoke of an inevitability of human progress and freedom?

It is from this new historical reality that my own intellectual inquiry into the future grounds of a Kantian program takes form. My question: Is it possible that the homogeneous process of abstraction and symbolic self-definition that Cassirer chronicled as part of the advance of Western civilization was itself only a contingent, and possibly localized, historical phenomenon? From the time of the first advances of the Greeks and then in bumps, jolts, and reversals, the inertial progress finally arrives at our own century—great advances in science, technology, and power, but now an international barbarism. Somewhere, we seem to have missed a fundamental dimension of human nature and thought. Indeed, it has been historically evident. The Western vision of progress earlier absorbed these unsettling realities as mere historical way stations. We can no longer be so self-assured. The inner structure of human thought that the Kantian tradition sought to lay out in terms of its formal structures as well as its presumed evolutionary progress seems in many ways to stand in the same relationship of Newtonianism with both Einsteinian and quantum physics, a special case of a much larger story.

In my four-volume study of the evolution of human intelligence, the problem addressed is this presumed homogeneity of the abstractive symbolic power of human beings.[23] Evidence is that the human species is a far more heterogeneous animal form than the philosophers of the West were ready to recognize. This cultural bias might be explained as part of a traditional historical naivete that subsequent events and time have brought to conscious awareness.

What would it mean, for example, if the evolutionary progress of culture, moving ever more toward abstractive symbolic structures of meaning and their supposed universalization, were only a Euro-

pean phenomenon, and a deeply flawed one at that? The suscepti-
bility to mythological behavior in a variety of symbolic modes—
political, aesthetic, philosophical, ethical, even technological—might
shake us from our universal subsumption of the human species under
the mandates of Kantian categories.

These intellectual questions derive from the human world of the
1990s and thus are properly different from Immanuel Kant's eighteenth-century purview, or Hermann Cohen's nineteenth-century
concerns, or even those of Ernst Cassirer, who brought us to the
threshold of our modern situation. If we are interested in the form-
giving propensities of humans as we act on the world around us,
shaping experience in terms of our inner intellectual and expressive
categories, not being mere passive recipients of external stimuli,
then, as philosophers we must cast our net wide and deep to delve
into the raw concreteness of empirical reality. By following science
carefully, we will be able to fulfill our philosophical responsibilities
to give meaning, coherence, and eventually confirmable truths concerning the human condition.[24]

The fruitfulness of Ernst Cassirer's Kantian program lies in this
never completed intellectual hegira to understand humans by understanding the principles and empirical grounds from which our behavior derives. The source of this knowledge lies with humans alone.
No exterior forces or criteria can help us avoid responsibility for
thought and action that lies mysteriously within our evolving nature.

Notes

1. Stephen Koerner, introduction to the English edition of Cassirer's
Kant's Life and Thought (1918) (New Haven, Conn.: Yale University Press,
1981), p. xiii.

2. Richard Cassirer (1868–1925), neurologist; Fritz Cassirer (1871–1926),
composer, conductor; Paul Cassirer (1871–1926), publisher, art dealer (Berlin
Secession); Bruno Cassirer (1872–1941), publisher (of Cassirer, H. Cohen,
and other philosophers); Kurt Goldstein (1878–1965), physician, aphasia researcher, psychologist.

3. Ernst Cassirer, *Leibniz's System in Seinem Wissenschaftlichen Grundlagen*
(Hildesheim: Georg Olms, 1962), "Einleitung: Descartes's Kritik der mathematischen und naturwissenschaftlichen Erkenntnis," Seite 3–102.

4. Ernst Cassirer, *Substance and Function* (1910), trans. W. C. Swabey and M. C. Swabey (1923) (New York: Dover, 1953).

5. Emile Meyerson, *Identity and Reality* (1908), trans. Kate Loewenberg (New York: Macmillan, 1930), pp. 229ff. Cited by Cassirer in *Substance and Form*, pp. 324–325.

6. *Einstein's Theory of Relativity*, trans. W. C. Swabey and M. C. Swabey, (1923) (New York: Dover, 1953).

7. Ibid., p. 355.

8. Ibid., pp. 455–456.

9. Ibid., p. 456.

10. Martin Buber, *Between Man and Man* (New York: Macmillan, 1965), p. 119ff.

11. *Einstein's Theory of Relativity*, p. 455.

12. *The Philosophy of Symbolic Forms*, vol. 4, trans. J. M. Krois, eds. Verene and Krois (New Haven, Conn.: Yale University Press, 1996).

13. Trans. O. T. Benfey, introduction by H. Margenau (New Haven, Conn.: Yale University Press, 1956).

14. *The Problem of Knowledge: Philosophy, Science, and History since Hegel*, trans. W. H. Woglom and C. Hendel (New Haven, Conn.: Yale University Press, 1950).

15. For citation in full, see p. 142.

16. *The Myth of the State*, p. 360.

17. Ibid., p. 375.

18. "Universal Natural History and Theory of the Heavens" (1755); "New Theory of Motion and Rest" (1758).

19. Seymour W. Itzkoff, *Ernst Cassirer: Scientific Knowledge and the Concept of Man* (Notre Dame, Ind.: University of Notre Dame Press, 1971) pp. 207–210.

20. Susanne Langer, *Mind: An Essay on Human Feeling*, 3 vols. (Baltimore, Md.: Johns Hopkins University Press, 1967, 1972, 1982).

21. Ibid., vol. 1, p. 291.

22. Ibid., p. 292.

23. Seymour W. Itzkoff: *The Form of Man: The Evolutionary Origins of Human Intelligence* (1983); *Triumph of the Intelligent: The Creation of Homo sapiens sapiens* (1985); *Why Humans Vary in Intelligence* (1987); *The Making of the Civilized Mind* (1990), all from Peter Lang International Publishers (New York).

24. See, for example, Seymour W. Itzkoff, *Human Intelligence and National Power: A Political Essay in Sociobiology* (New York: Peter Lang, 1991); *The Road to Equality, Evolution and Social Reality* (Westport, Conn.: Praeger Publishers, 1992); *The Decline of Intelligence in America: A Strategy for National Renewal* (Westport, Conn.: Praeger Publishers, 1994).

Preface to the First Edition

The purpose of this study of Ernst Cassirer's philosophy is twofold. The first is principally a matter of exegesis, to delineate from Cassirer's own writings the historical and systematic sources for his intellectual position. The second is to examine the implications of this critico-idealistic philosophy for a theory of man and discursive knowledge, two pervasive themes in Cassirer's neo-Kantianism.

One of the problems in assaying the significance of Cassirer's philosophy—due in part to the difficulties in translating his philosophical style—is the fact that his epistemological position is diffused in an implicit manner throughout his writings rather than explicitly stated and argued. The major reason for this approach is Cassirer's preoccupation with intellectual history. We have in this sensitivity to the concerns of the past a unique instance of a creative philosopher who, rather than enunciating his own radical philosophical approach, joined his insights to the main lines of thought of our modern era. Thus, though Cassirer, after the first early volumes of *Das Erkenntnisproblem*, turned from the study of intellectual history to more synthetic philosophical pursuits of the nature of scientific knowledge and thought, he never abandoned the historical point of view. Rather, he attempted to show that Marburg neo-Kantianism as a philosophy was in agreement with scientists' discoveries about the theoretical structure of their own disciplines. And in fact as a philosophy neo-Kantianism was completing and elaborating those insights that the earlier thinkers intuited only inchoately.

The historical dimension thus interpenetrates the entire corpus of Cassirer's mature philosophical writings. It is an ever present reminder that in Cassirer's view the great thinkers of the past were not half-seeing purveyors of pseudo-truths, objective knowledge forever eluding them. To Cassirer any particular state of contemporary knowledge was always constituted of the problems of the past reinterpreted in the light of the contextual circumstances of the present. He was aware that his own philosophy was also conditioned by the opportunities and limitations afforded by contemporary science. It will itself be judged by history.

In delineating the direction of Cassirer's philosophical development in a manner true to the character of his method and content, we will of necessity refer to his own interpretation of the history and evolution of scientific knowledge. To this end the earliest chapters (1, 2, 3) derive from his mature writings that perspective of the older tradition which Cassirer utilized to lend meaning to his contemporary investigations. It is in this sense that Cassirer's epistemological research flows out of and is organically related to the philosophical tradition that began in the seventeenth century and reached a first climax in the Kantian synthesis.

Cassirer's application of this Kantian tradition for an analysis of contemporary scientific and culture theory is discussed in the second section of this study (4, 5, 6). The philosophical message is of course Cassirer's. What is stressed in this presentation is not merely a condensation or systematization of his analyses and conclusions in the various theoretical domains. An explanation is offered as to the inner logic of enquiry which guided Cassirer in his study of the various forms of knowledge and which led him to turn for an understanding of the meaning of man's theoretical powers from the all-encompassing abstractive realm of the physical sciences to culture theory, language, neurology and psychopathology, and finally to evolutionary biology.

The final part of the book (7, 8, 9) deals briefly with several problems left to us by Cassirer, i.e., human nature and man's symbolic propensities, the meaning and direction of man's cultural tradition, and finally the status of Cassirer's instrumentalist and antiobjectivistic views of discursive knowledge in the light of recent views in epistemology and the philosophy of science. These chapters of course do not pretend to provide a definitive elaboration of Cassirer's position, but merely suggest some questions which Cassirer himself might have confronted had he had the opportunity.

Another aspect of this final part should be mentioned in that it attempts to fulfill a methodological stricture which was uniquely characteristic of Cassirer's work—the utilization wherever possible of evidence from the various empirical disciplines to muster support for the philosophical claims he made. There are some, as I have noted in this study, who because of Cassirer's supposed Germanic philosophical style, associate him with traditional continental schools of metaphysics. There is no support for this prejudice. In naturalistic outlook, intent, and achievement, in openness to factual evidence, firsthand knowledge of the physical and cultural sciences, Cassirer is a contemporary.

My interest in the Kantian philosophical tradition began with a study, written under the direction of Ernest Nagel of Columbia University, of the scientific philosophies of Pierre Duhem and the French neo-Kantian Emile Meyerson. Eventually this led to Cassirer and an abiding fascination with his writings. I am indebted to Smith College for providing the conditions favorable for the completion of this study. My wife has provided vigorous stylistic criticism for which I am appreciative. To my colleague Leonard Baskin I extend an expression of gratitude for his masterful drawing of Ernst Cassirer. Finally, I would like to thank Miss Ann Rice of the University of Notre Dame Press for her judicious editorial assistance.

I

Cassirer's Kantianism:

The Historic Issue from Galileo to Hume

KANTIANISM IS THE OFFSPRING OF A PHILOSOPHICAL
crisis. The great rush of intellectual confidence and achievement
which began with the new physical science in the seventeenth
century led in the eighteenth to a philosophical standoff between
the rationalist and empiricist schools. Immanuel Kant's great synthesis resolved the dilemma. But it too was soon under attack for
alleged philosophical inadequacies and for a time was eclipsed.
The currents issuing out of the Kantian doctrine however were
not to be dammed up. They would appear over and over again
in new contexts and with regard to new issues.

Ernst Cassirer determined to learn from the historic state of
affairs. His writings reflect a deep and continuing sensitivity to
the fact that every philosophical position is itself eventually
judged as history. To understand Cassirer and his neo-Kantianism one must also understand the role that Kant plays as inheritor and synthesizer of the philosophical tradition. We will follow
with Cassirer the intellectual dialogue that attended the birth of
modern science and which dominated the minds of the European
intellectual community for over a hundred years.

Science too had been the product of an earlier epistemological
crisis. It was an attenuated crisis in that the older truths had

already lost much of their vitality. An awareness of the power of number, observation, and experiment had gradually infiltrated the consciousness of the educated until the day that Galileo confronted them with the startling reality of a philosophical *fait accompli*.

1

The intellectual revolution that was set in motion by Galileo unhinged a stable vision of world and man hitherto an integral part of the heritage of both humble and powerful. That ordered universe was perceived through intellect and experienced in the nature of society. It thus fulfilled one of the most basic cognitive needs of man, that of interpreting his own personal destiny in the universe in the light of his intellectual traditions and in accordance with the expectations of the society around him.

In this Aristotelian-Christian view of things, the hierarchy of substances placed unformed matter at the base of our schema of values and reality. At the apex of the matter-form structure existed God—pure form. God as *telos*, as first efficient cause, as object of aspiration, impelled the universe into motion—into the actualization of its potentialities. Man too, soul and body, form and matter aspired to knowledge of God as well as the experiential world. His destiny, fixed in its appointed tier in this universe of substances, was determined and fulfilled in the process of becoming rational and arriving at an intellectual vision of the world, himself, and God. But because his substance was anchored in this eternal, soaringly Gothic embodiment of reality his potential for evolutionary change was nonexistent. This was an active, mobile, energetic, and even natural world, yet, like an army forever marking time, it was going nowhere.

Intellectually this was satisfying, and perhaps experientially too. Both supernatural and natural qualities in man would be accounted for in thought and life. Sensory experience, Thomas Aquinas told us, was initiatory in bringing the nature of reality to the attention of our natural capacities. In theory man could rise from the limitations of sensation to behold in its intellectual glory the ultimate nature of the universe of which he himself was a part. Sensory experience as actualized by the intellect—itself part of the greater and eternal wellspring of reason—was also both natural and eternal. Knowledge of the workings of matter as well as of the existence of God could be achieved by man. Both *sapientia*—the highest forms of knowledge, and thus of the supernatural—and *scientia*—knowledge of the material world— Cassirer concludes, are completely accounted for and united within a systematic and coherent philosophical explanation of reality.[1]

But permanence and coherence were not enough to enable this vision to endure. The truth of this system could exist only as long as that particular intellectual esthetic quality exemplified in the soaring monuments of the Gothic cathedral builders re- mained consistent with the hardly changing vistas of the world about. The inherent fragility of the Thomistic synthesis came to light only slowly. The creative surges of Renaissance life and thought finally upset the socio-political context of the medieval world, while the reforms which came out of neo-Pythagorean and neo-Platonic mathematical mysticism undermined the Aris- totelian metaphors.[2] The work of Copernicus, Brahe, and Kepler provided the foundation for the final revocation brought about by Galileo.[3]

An important step was taken, Cassirer states, when Galileo exclaims that the universe can be understood in mathematical terms alone.[4] To Galileo a new ontological light is shed upon the nature of reality by the laws of mathematics. Not only are the motions of bodies to be understood through these mathematical principles, but experimental proof can be derived from these self- same laws, to be the touchstone by which they are accepted as true. Reality and truth are thus seen to reside in the mathemati- cal principles. The "real things" of our universe to be substituted for the hierarchy of substances of Aristotelianism are the Democ- ritean atoms, abstracted from the geometrical bodies moving in mathematical space and time.[5] This world of atoms is nothing but the abstract representation of physical reality, insofar as nothing is retained in its concept but pure determinations of magnitude. Thus all reality reduces to these mathematical bodies in motion.

Galileo holds that the epistemological status of his law of inertia would be maintained even if no experimental evidence could be acquired for it.[6] This is, he argues, because our knowledge of nature depends only upon the theoretical possibilities inherent in the new geometrical method. The metaphysical reification of mathematics reflects Galileo's reaction to the sterile formalism inherent in the hierarchy of qualitatively unique substances in Aristotelianism. He reacts by substituting its metaphysical antithesis—the qualitatively neutral mathematical Platonism of Kepler and the north.

The crucial metaphysical crossroad of this philosophical revolution occurred when Galileo turned to assay the role and status of man in a universe dominated by the geometrical method of mathematical bodies in perpetual motion. The truth qualities of this new method, its invariability and permanence, led him to postulate that the source of the traditional error of the academic method lay in the fact that its determinations were developed from common sense. The instability and flux of sensation from which knowledge traditionally is derived were further evidence of the deficiencies of this method. Sense perception can never give us an indication of the nature of the real. The conventionality, the arbitrariness, and the accidental quality of the senses are similar to language and names—a rich source of the materials of scholastic disputation. These are all of a lesser epistemological level of truth, an area of secondary qualities.

But how does man attain the truth of mathematical principles if not from experience, whose channel of knowledge is provided by sensation?[7] Thomas's naturalism led him to postulate sensation as the starting point of all knowledge. Galileo, Cassirer noted, could not admit sensation into his system. "None of these concepts is derived from experience; the mind takes them 'from itself' in order to apply them to sense preceptions."[8] Man then must in some manner produce the concepts of mathematics, then apply them to sense experience to verify and correct our mathematical principles. The objective world, wherein lie knowledge and truth, objectivity, and necessity, is in this way sundered from the contextual perspectives and autonomy of organic man.

The break with the past was now complete. The ecclesiastical bonds with which the scholastics attempted to encompass all human intellectual activity were broken and discarded. The organic hierarchical unity of natural and supernatural worlds within which man played an integral if mediate role in this architecture of substances was now dissolved. Instead, a new division between the real objects of the world and the illusory was offered. The real was equated with those characteristics of objects affording the possibility of reduction to the mathematics of abstracted atomistic bodies moving in geometrical space and reversible time. The descriptions of experience—feeling, common-sense objects, sensations—were now to be only secondary qualities of things, mere names. They would be accidental and subject to random changes, contrasting with special vividness to the clarity and utter necessity of the ontologically superior world of mathematical principles.[9] We could ascribe truth and knowledge only to the laws and theorems of mathematics within this new physical science.[10] Galileo made these principles available to the mind of man but only through an opaquely defined process of thought. The objects of truth and knowledge were not empirical things but mathematical concepts coinciding with Galileo's postulation of an ontologically higher reality.[11]

If man lived completely within the world of common-sense processes, events, and things, he could not be encompassed by the mathematical laws of bodies in motion. Therefore he had to be relegated to the nether world. Human individuality was no longer an important epistemological factor. It was merely the center for the reception of perceptions and bundles of sensations. At the same time Galileo granted man the enigmatic privilege of producing almost spontaneously from his mind those same mathematical principles from which the laws of nature are constructed.

The power of this mathematical method to order experience and to guide us in future experiments to control our physical environment was enough to persuade Galileo to discard what had been an acceptable intellectual structure of experience. Obvious distinctions between the world of things, animate and inanimate, which had been accounted for in a manner satisfactory to common sense by the Aristotelians, were to be abandoned. Man is sundered from his comfortable relatedness in the order of the real. His essence or nature is now split between his role of conceiving the mathematical laws and concepts of the new science—which truly describes reality—and the subordinate and secondary epistemological activities of sensing, verbalizing, and experiencing within the world of common-sense behavior.

Galileo, Cassirer stated, introduced sharp distinctions between things that had been confused and he united natural phenomena that hitherto had been regarded as disparate and heterogeneous.[12] The inheritors of his philosophical revolution are thus bequeathed a problem with which they will wrestle for two hundred years. Not until the concepts of knowledge, man, and reality were reinterpreted in the Kantian synthesis was a temporary solution found. It was formulated within the context of the Newtonian astronomical vision and the resurgent Rousseauian and German idealism.[13]

2

Perhaps it was Descartes's traditional humanistic Jesuit training which sensitized him to the paradoxes in Galileo's ontological sundering of man and matter. The laws which came "full blown" from the mind of man and which described and revealed the reality of matter in motion to Galileo seemed to partake of an epistemological limbo. Descartes's first concern, Cassirer believed, was with this problem and with the methodological issue that was created through man's epistemological investigation of the real world. The *cogito ergo sum* was an attempt to reestablish mind as a metaphysical equal in the universe without negating or compromising the status of the mathematical primary qualities as against the ephemeral character of sensations and secondary qualities in general. Descartes saw Galileo's vagueness in ascribing the sources of these laws and ideas, notwithstanding the detailed methodology of scientific proof in the hypothetical method, as the key problem in the reinstatement of mind into the ontologically real universe. In fact, only through establishing mind as a fully equal partner with these mathematical laws of science could we validate the latter epistemologically.

Added to this was Descartes's claim that the mathematical view of matter in motion was incompatible with Galileo's hypothetical postulation of atomism.[14] Descartes's replacement of a substantial materialism and the substitution through geometry of *res extensa* and *res cogitans*, dual realities of which men can have knowledge, pushed Galileo's original train of thought to its logical conclusions. The Cartesian material substance is not only defined through geometry, but its very reality is permeated with and constituted of mathematics.[15]

The equality of substances in Descartes's postulation is not unconditioned as compared with the status of those principles underlying Galilean science. God is no longer merely the knower of the mathematical laws of bodies in motion or the vague first cause of all motions. Surely the truths of nature which man learns are the same as those which God knows. In Galileo's conception, we can know them as well as God does. Descartes was more timid; he postulated the ontological dependency of all knowledge upon God and thus the conditional character of the dual substances.[16] The highest metaphysical substance and thus the supreme reality of which we can achieve knowledge is God.[17]

Descartes retained the identity of the object of knowledge with the ultimate objects of reality. Extended substance and the geometrical symbolism by which substance is explained are metaphysically identical. But this identity is given meaning or achieved by carefully differentiating between pure method or concept and substance itself.[18] What had occurred, Cassirer explained, was the transformation of what in Galileo had been a form of conceptual realism into a true substantial realism. Where Galileo did not clearly delineate the method by which mind came to its mathematical ideas of ultimate reality, but merely postulated the discovery of the laws of mathematical bodies in motion by mental inspection, Descartes felt the need for a separate substantial category—mind—to meet these methodological shortcomings. Mind, in a sense, is still separated from the human context. It is the methodological handmaiden for the achievement of absolute knowledge of the external realities of geometrical nature (*res extensa*).

It is here, Cassirer stated, that the separation in tradition between the lines of empiricism and rationalism occurred. In Galileo the union of rational and empirical methods was in a state of becoming.[19] His philosophical writings reflect the developmental nature of his thought. By the time the tradition fell into Descartes's hands, the union was established. But Descartes's needs were different. They reflected a personal methodological involvement in ascertaining the source of the knowledge of this new science. This methodological quest resulted in the reification of a new substance co-equal to mathematical substance.[20] But now the emphasis was on ideas in both realms—nature and mind. The experiential and experimental orientation of Galilean science was abandoned as the philosophical scene switched from its revolutionary origins in Italy amidst the debris of the now finally discarded Platonic and Aristotelian traditions.[21] The reconciliation of the real substances given in mathematical physics and in introspective mentalism by the dualistic parallelism of Descartes now engaged the continental mind.

3

With Spinoza and Leibniz the rationalistic tradition continued its involvement with the implications of physical science for man.

No essentially new ground was broken by either of these figures. Rather, a dialogue ensued, with each philosopher noting the logical implications of the other and drawing these implications to their natural ends.

In Spinoza the dualism of mind and extended substance under the overseership of God was resolved in an all-encompassing naturalistic monism. "The general laws of nature which govern and determine all phenomena are nothing but the eternal decrees of God which always entail eternal truth and necessity."[22] Characteristics of God are determined by the laws of nature and are in fact synonymous with these laws. If therefore, "things could have been of another nature or determined in another manner for action so that the order of nature were different, therefore, also the nature of God could be different than it is now."[23]

Thought and knowledge are equated with reality. Thought creates the implicatory truths of geometry, which encompass all of reality. There is no complementary or parallel substantial domain (*res extensa*). Extended substance is itself thought, the universal substance, God or nature, depending on one's outlook.[24] God is never in any one thing or set of attributes. He is the intelligible order that makes all rational.

Spinoza's God-intoxicated pantheism constituted a logical *tour de force*, Cassirer noted, in drawing the implications of this self-subsistent structure of eternal goemetrical forms for a vision of the universe.[25] Human autonomy, temporality, sensations, and change, however, were relegated to an intellectual limbo.

Leibniz's monadology is a step backwards from the rationalist identification of mathematical law with real substance. Instead, a dynamic, organic plurality of substances is postulated, each monad reflecting a particular and partial perspective of the totality of knowledge.[26] Mathematics in this view becomes an organ of knowledge, not a quasi-substantial entity defining and interpenetrating a substantial being. Leibniz recognized that without the use of mathematical signs we could not develop our views concerning the mechanical character of nature. But he was the first to develop explicitly the theme that these mathematical laws are not ontological in nature. In Cassirer's terms concerning Leibniz's use of mathematics: "[It] is an instrument, by means of which this content develops and fully defines itself."[27] The special symbolic and conventional character of mathematics is part of Leibniz's idealization of the laws of nature as phenomenal, behind which he was able to introduce the reality of his monadology.[28]

As part of this program, Leibniz attacked the Cartesian conception of material substance as extension. By the standards which Descartes himself laid down—intelligibility and clarity—Leibniz claimed that extension is an illegitimate concept, permeated as it is, in its geometrical representation, by a host of intuitive and imaginative characteristics.[29] He wished to replace extension with the pure idea of force, itself reducible to mathematical principle.[30] Science would thus be guided toward a unity of ideal concepts. In this next phase, all empirical existence and empirical events would be related and ordered in accordance with the demands of intelligible truths. In Descartes's words, "la vérité étant une même chose avec le être" (*Meditations*, vol. V).

The problems of sensation and the origin of knowledge reenter continental philosophy in Leibniz, reflecting the growing uneasiness in England with this rationalist tradition. Sensations, Leibniz claimed, merely bestir ideas in our minds. Knowledge may originate in this activation but it can never pass as knowledge because there is no necessity involved in sensation.[31]

Leibniz resolved the problem of God by placing Him in a transcendent ideal position as the highest monad, the most complete perspective of the universe. The interrelationship of all the monads being determined at the creation of the universe, God becomes the first cause of all events which now interact through this preestablished harmony.

Though dependent upon God, the truths and ideas which a Cartesian came to know through the dual substances presented as true a rendering of reality as those which God Himself knew. Leibniz structured his metaphysical hierarchy in another way. Man is a monad situated midway in the hierarchy of monads, much as he stood in the Thomist synthesis.[32] Each monad reflects the rationality and universality of the universe in its vision, in a line ascending to God.[33] True rationality, the universal knowledge of all phenomenal relationships, is encompassed by the creative omnipotence of God. Mathematics is an instrument by means of which the phenomenal content develops and fully defines itself. Thus knowledge of the universe of objects, achieved through "mere symbols," becomes blind knowledge when measured by intuition, "the true sight of the idea."[34] The symbol, Leibniz stated, is limited and finite in contradistinction to the ideal of the perfect, archetypal, and divine intellect. Thus the symbolic character of knowledge of the world of nature acquires a negative connotation in comparison with the higher metaphysical knowledge given in the intuitive idea and revealed in the system of monads.

It was Leibniz's symbolic and idealistic conception of knowledge that provoked the famous correspondence with Clarke, Newton's protagonist, concerning the nature of space, time, and gravity.[35] Leibniz argued against the realistic and absolutistic views of the Newtonians. Space and time cannot be considered as real and absolute.[36] Space is no more than the ideal condition of order in coexistence and time the condition of order in succession. Thus though the Newtonian laws are eternal truths, they are purely ideal. They do not constitute the contents of reality.[37] On examining this phase of Leibniz's philosophy, Cassirer labeled Leibniz as the precursor of much of the modern symbolic conception of mathematics and logic as well as physics.[38] Modern science has adopted Leibniz's views on the logical status of time and space in a manner similar to its acceptance of Descartes's vision of the extendedness and fulness of space. Cassirer postulated that Descartes anticipated the field theory of space as consummated within Einstein's general theory of relativity.[39]

In general, Leibniz's continental rationalism marked the transition from strongly realistic epistemological tendencies to a more idealistic perspective of knowledge, anticipating both Kant and Hegel. "What we call the contemplation of the nature of things is very often nothing but the knowledge of the nature of our own minds and of those innate ideas which one does not need to seek outside" (*Nouveaux Essais*, bk. 1, chap. 1, sec. 21).

Yet those ideas are there only because of the all-seeing power of the highest monad—God. Leibniz disregarded the human dimension in the making of this albeit symbolic knowledge. It is merely an adjunct to a superior power out of which reason itself flows. Leibniz did not derogate knowledge; he merely tried to shift its status and source to a theological and hierarchical context rather than the naturalistic base that had been established up to that time. "There is nothing in heaven or on earth, no mystery in religion, no secret in nature which can defy the power and efforts of reason."[40]

4

With Locke the empiricist tradition was turned to the problem of the origin and derivation of knowledge from experience. Thomas Hobbes, system builder yet empiricist precursor of Locke, was not especially interested in the problem of the derivation of knowledge. This problem, which the new science had precipitated upon the intellectual scene, was in essence avoided by his imputation of the source of knowledge to the Galilean bodies in motion. We could receive the sensations which these bodies caused only through a process of efficient causation within our own bodies.[41] Once received, these sensations were ordered into concepts through signs which were the arithmetic of science. Hobbes never really grappled with the nature of the efficient causation by which bodies caused sensations to appear to us. He also left unanswered the method by which names—our only form of ordered knowledge—would reveal the reality of the bodies in motion to assure us that our science truly described the action of a body in motion taking place beyond our sensory organs. The causal connections between geometrical bodies and sensations, the transformation of sensation into concepts, and finally the agreement of concept with reality were left unexplored.

Cassirer saw the empirical tradition as being refocused in the thinking of John Locke. Locke attempted to reestablish the independence and autonomy of experience as a source of knowledge.[42] Locke was a rationalist in that he believed in the possibility of possessing knowledge of the external world. But the important consideration for him was that knowledge must stand on its own feet. On the continent the new scientific method had revealed a new form of knowledge, one which had resulted in the coupling, if not in the complete identification, of knowledge with reality. In each of the great systems this philosophical process of thought had led inevitably to the question of a genetic source of this knowledge. The clear and distinct ideas which were postulated could not be grounded in sense perception. The necessary knowledge provided by these ideas had nothing in common with the transitory character of sensation. Whatever the solution offered to the problem of the origin of knowledge, whether a Cartesian dualism subsidiary to a transcendental God, a naturalistic monism (Spinoza), or a pluralistic organicism (Leibniz), it could not have satisfied the empirical and critical temper of Locke's mind.

Locke offered us another solution. Rather than focus his analytical tools on the object of knowledge or on the mathematical deductive necessity inherent in the ideas which had transfixed the continentals, he sought the genetic origins of knowledge. Psychology would replace metaphysics. He attempted to analyze critically the process of experience. And by putting man (albeit an abstracted vision of man) back into the center of this process, Locke effected a compromise which allowed an empirical interpretation of experience that did not attempt a penetration of the inwardness of reality, yet, Cassirer felt, still maintained the tenets implied in the scientific world view, now an established and unimpeachable wellspring of all knowledge.[43] Not since Galileo had the "knowledge problem" been developed in a way which combined the rational with the experiential. Of course the prime question was whether or not Locke would be successful.

As a consequence of Locke's analysis of experience concerning the origin of knowledge, we could logically postulate the existence of simple ideas furnished to our mind through cores of sensation. The mind is furnished with these simple ideas from an external source, presumably, said Locke (here in agreement with Hobbes), by the objects capable of producing these ideas. Since the essence of things is unknowable, at most we can interpret the external objects as an aggregate of perceptions. Along with the object and inseparable from it are those qualities of solidity, extension, figure, motion, and number which in science define body. The primary qualities also carry with them the secondary qualities of color, sound, taste, etc. The objects of science thus take epistemological precedence over the disorganized, haphazard qualities of common sense. As a problem, substance, that ever elusive quality of the rationalist tradition, is shelved, for it is seen as an aggregate of properties alone. It does not have to call into use any inner necessity or bond, something around which these properties cohere.[44]

The problem of achieving determinate knowledge is resolved by the postulation of an inner tendency in our mind that would allow sense data to embody a trend toward universality. This concept—the combination and separation of simple ideas—issues into three general types of complex ideas, which Locke called modes (abstract triangles, number), substances (material, mental, God), and relations (cause and effect). Ontological reality is given only to those simple cores of sensation, the building bricks of knowledge and experience, which form the simple ideas. Locke was emphatic about this ontological separation of concept and object: "We may refine our intellectual instruments and advance from sense perception to universal concepts but we can never establish anything more than a coexistence of ideas and sense perceptions."[45] The coexistence of our ideas then is the furthest we can advance through Locke's epistemology toward the necessity which physical theory demands from philosophy.

In regaining for human experience its central role in the production of knowledge, Cassirer argued, Locke was forced to surrender the rationalist attempt to encompass reality within the logical structure of the mathematical and physical concept.[46] We must be content with more modest goals. We can know only the given, the simple ideas which enter as phantom reflections of a world of bodies whose substance we can never know. Like Hobbes, Locke abandoned the attempt to penetrate reality. But unlike Hobbes, Locke, in his analysis of the comparison—i.e., the agreement and/or disagreement—of these simple ideas and in his tracing of their development into complex ideas, the notions, does give us the means for a world of knowledge.

In its general aim to provide a solid epistemological base for the new science, Locke's rationalism concludes by approaching the Cartesian postulation of clear and distinct ideas. The crucial difference, even considering the superficial similarity of goals in the two philosophies, lies in the metaphysical character of these theories of knowledge. Descartes's clear and distinct criteria are ultimately grounded in the mind of God, giving us a correspondence theory of knowledge. Locke's postulation of the agreement or disagreement of ideas leaves the issue of knowledge to logic and experience, giving us a coherence theory of knowledge. Yet when all else has been considered, we are forever thrust back into the problem of experience as we concern ourselves with the source and origin of knowledge. If there are inherent difficulties in the epistemology of both rationalism and empiricism, they occur because, as Cassirer stated, both rationalism and empiricism seek to explain the metamorphosis by which the phenomenon develops from a mere datum of consciousness into a content of reality of the outside world. They fail because neither theory can explain that a phenomenon already points to an 'objective reality.'[47]

5

The quest for the object of knowledge now shifts to this problem of the nature of the simple ideas. Condillac, in France, presaging the Enlightenment, was soon to analyze the simple ideas which to Locke had been original and irreducible wholes—the data of sight, hearing, motion, taste, observing, willing—into their exact sensational components. The tenuous balance between sensation and reason in Locke's neat scheme was soon to feel the incisive thrust of Berkeley's logic.

The basic issue on which Berkeley pressed Locke's system, Cassirer felt, was the latter's sensationalistic rationalism. If, Berkeley maintained, we claim with Locke that the only reality we can know are the individually determined concrete sensations—simple ideas—then anything which goes beyond sensation, such as reflection, complex ideas, objects of the external world, etc., is fiction. If the foundation of all truth be in simple sense data, only more fictions arise as soon as the foundation is destroyed.[48] Under Berkeley's destructive logic the entire rationalistic psychology which Locke had erected to make the scientific world view plausible began to crumble. How such a rationalistic process could be created merely from the reception of sensation was Locke's primary difficulty. The very concept of a simple sensation or simple idea in Locke's use was also questioned by Berkeley. In Berkeley's use of simple sensation it soon became clear that the corresponding term in Locke denoted something which was hardly simple or irreducible when subject to analysis. Locke's reinstatement of secondary qualities likewise brought about the conclusion that primary qualities, such as shape, number, extension, etc., which adhere to all bodies, were also fictions of the mind.

Not only is reality exhausted by the original perception; it now becomes the primary datum of knowledge. It is here that the dogmatic purposes for which Berkeley initially launched his attack become clear. If we reject the premise that any rational determination arises from sense perceptions, if we eliminate those primary qualities and the objects for which they supposedly stand and which supposedly cause the sensations we receive, we are left with the fact of sensation as a completely isolated phenomenon. But, Berkeley explained, this is not an insoluble dilemma. Perception, or simple sensation, is the original language in which God speaks to man. Rather than God buttressing the mathematical and physical conception of the world, as Descartes postulated, Berkeley used Him to validate an idealistic psychology of perception. As Cassirer pointed out, Berkeley summoned inner experience to battle against outer experience: psychology against physics.[49] Berkeley's antagonism to Newtonian science as an important factor in the undermining of faith is a basic component of the unique character of his philosophical dogmatism. He is probably the only philosopher of the time whose interests in epistemology turn away from the scientific world view.[50] Concepts such as space, time, and matter become fictions of the mind because they are not reducible to a simple perception.[51] These concepts are pure names and as such are conventions.

But note the difference from Galileo's similar evaluation of names as conventional constructs. With Galileo, because reality was equated with the physico-mathematical structure of knowledge, names were equated with sensation. Berkeley considered reality to consist of pure sensations and names to be attributed to higher abstractions, such as matter and space. If matter is a fiction, it is but a short logical step to something no empiricist had as yet entertained—denying the existence of an external world beyond our impressions. Berkeley did not go quite that far; it was enough for him to note how tenuous the traditional epistemological assumptions of this fact have been.

This is the turning point in Berkeley's more critical temper. Cassirer felt that Berkeley was too much a child of the Newtonian revolution to deny the existence of the powerful and useful intellectual constructs by which individuals order their experience. One does not have to look far even in that environment to note that there is more meaning and significance to our world than pure sense perception would lead us to believe. Thus in order to save his philosophy, Berkeley appended certain auxiliary relationships to sense perception. He did not deny, for example, that there is such a thing as space, even as a mental fiction. To what source can be attributed the creation of this concept? What is its epistemological status? In Berkeley's terms a concept such as space becomes a relation between individual sense data. To the initial content of the mind, the perceptions of sensation, habit, and recurrence constitute a representative function, achieved when consciousness passes from one type of sensation to the other. This, Berkeley stated emphatically, is not a rational process, merely a repetitive one, i.e., the constant accompaniment of certain particular sense data. The concept is resurrected in still another manner.

To Berkeley the epistemological archdemon of rationalism was the general idea, not the concept; the latter was interpreted in the nominalistic sense as the unity in a rule. The general idea by which a triangle becomes the archetype of all triangles—isosceles, scalene, or equilateral—is, however, an inadmissible construct. By way of emendation, Berkeley stated that though there can be only one triangle perceived through the senses, by the unity contained in an arbitrary rule we may conceive the regularities of the rule as applying to other triangles.[52] Thus in spite of arguing vigorously against physical science, Berkeley came to terms with the problem of the physical concept and made a place for it within his epistemology, albeit a secondary position.

This strong-minded attempt to eliminate the dualism of sensation and reflection by reducing reality to sensations and sense to a few ordering rules had still left serious doubts as to the success of the enterprise. Faced with the logical consequences of his nominalistic sensationalism, Berkeley retreated by way of the readmission of order, rules of repetition, etc. He thus implicitly accepted the existence of a degree of rationality and logic, without which all synthetic knowledge would disappear in an avalanche of confused sensations.

Indeed, Cassirer stated, in Berkeley's final years, possibly pursued by this rationalistic specter, he postulated in his *Siris* that concepts of bodies are nothing other than the sensuous sign language in which an all-embracing spirit communicates itself to one finite spirit.[53] God's thoughts are the source of our physical world. In fact the real world recedes from the immediacy of perception and like Platonic archetypes is equated with the thought of God, which reaches man in its particularity in His knowledge of the physical world. It is sad to note this weak and vapid conclusion to a life of as tough-minded an empiricism as one could expect at any time in the history of thought.

6

With Hume the empiricist reaction reaches its acme. From Berkeley's devastating critique we come to an all-pervasive scepticism concerning the rational basis of knowledge as an auxiliary buttress to the scientific world view. In Hume the psychological critique is completed. Adopting most of Berkeley's epistemological empiricism, Hume merely emphasizes its sensationalism. Objectivity is again grounded in the simple impression. But Hume avoided Berkeley's God as an explanation of the cause of these impressions. He was a trenchant critic of any so-called proofs of God's existence. This critical attitude derived both from his secular empiricist proclivities and his psychological scepticism. Man's need for God, rather than a manifestation of objective reality, is a subjective manifestation of certain fundamental and original instincts of human nature.[54] The origin of any single impression can only be attributed to unknown causes existing external to us. Our ideas are a form of internal sensation and they too are inexplicable in origin. The bond which unites our ideas is just as incomprehensible as that which unites external objects. The self in general is to be considered as no more than a bundle of sensations, not in any way as a substance, as in Locke, stated Cassirer.[55] These sensations, together with such conventional ordering forms as similarity, imagination, and resemblance, give impressions the appearance of unity.[56]

This last statement suggested to Cassirer the difficulty faced by Hume, like Berkeley before him: to account for all conceptual structures which we experience in the process of knowing. Hume attempted to resolve the issue by postulating concepts such as consistency and coherence, which he stated to be fictions or delusions of the imagination stemming from universal psychological laws, therefore with no epistemological or logical status.[57]

Hume wrote that "the 'thing' is not merely a bundle of individual perceptions, but that the idea of an object identical with itself first arises through the concepts of constancy and coherence . . . . This concept is a mere fiction—a delusion of the imagination to which it must succumb in accordance with universal psychological laws but to which we may attribute no objective-logical value."[58] Causality, today an example of a rule or maxim of science, is thought not to be a necessary concept by Hume, and is hypothesized to issue from the habit of noticing the constant conjunction of sensory impressions in consciousness with great enough frequency to allow the imagination to hypostatize spatial coexistence or temporal relation into a causal relation. This position is considered by Cassirer to exemplify the remarkable similarity between primitive predications of causality and those of Hume. It caused Cassirer to view Hume's position as a modernistic philosophical version of the workings of the primitive mythic mentality.[59]

Hume was far from being an enemy of science; but he was certainly the greatest philosophical sceptic of them all. We can have two types of knowledge he tells us. The first, involving relations of ideas—algebra and geometry—is intuitive, demonstrable, and certain, much as Leibniz conceived the knowledge of signs by human beings. The second concerns matters of fact which derive from the relation of cause and effect, which we get secondarily from our impressions of sense data. The most we can expect from this is probable knowledge.

In a sense Hume's philosophy, to Cassirer, is the explication of an incomprehensible riddle. The only presupposition or reality he starts with—the particulate sensations—denies any status to our ideas. Ideas are the vague and probabilistic creations of our imaginative tendencies.[60] But even in the distinctions which Hume made between sensations and ideas, the elements of our thought, we are left with this residuum of incomprehensibility. For, as Harald Höffding once pointed out, "to comprehend is to unite, to find coherence," and this Hume said, we cannot accept as real.

2

The Critical Kant

THE DIALOGUE BETWEEN RATIONALISM AND EMPIRI-
cism over the nature and status of the elements of knowledge
endured for almost two centuries. By the middle of the eighteenth century the situation had reached a precarious state. The
later rationalistic metaphysical systems represented by Leibniz
and Wolfe had been subjected to the rigors of Hume's analysis.
Utilizing almost mathematical deductive methods inspired by the
achievements of scientific thought, the rationalists had assumed
that reality could be penetrated and known by a structure of
ideas contributed by *a priori* principles inherent in thought itself.
These clear and distinct ideas, through which all of reality, from
common sense through the nature of God, could be known, did
not arise in experience; nor were they endowed with a sensory
character. Humean naturalism had unmasked these claims as
chimerical.

In the rather provincial atmosphere of Königsberg, Kant, for-
tunately perhaps, had been spared acquaintance with the arid
philosophical interpretations of the historical tradition. It is possible that this disinterest merely mirrored the typical Enlightenment attitudes toward the world of past and present. The power
of the new science to reveal to man a vast area of knowledge
and the great capacity of the human mind, unhindered by any
theological dogmatisms, to penetrate the distant recesses of the

universe, were sufficient to eradicate much feeling of historical dependency in the leading minds of this era.

Kant's initiation to epistemological thought came from two sources. First was his well-known involvement with Newtonian philosophy.[1] Indeed, his creative contribution to cosmology developed out of his concern with Newtonian physics.[2] On the other hand, philosophical influences on Kant were mostly negative until the arrival of Hume. Kant was relatively uninterested in and thus ignorant of the writings of Descartes and Leibniz, as well as of the empiricists.[3] This was perhaps due to the contemporary disintegration of philosophical creativity in this era after Wolfe. In the field of epistemology the psychological method of empiricism was being used to buttress a teleological dogmatism in the spirit of Leibniz's "best of all possible worlds."[4]

Kant, like Galileo before him, wrestled with the problem and the necessity for integrating these two diverse intellectual strands of science and philosophy. The status of knowledge, mind, and sensation hinged on the solution. When in about 1763 Hume's writings became available in translation and, as Kant confessed in the preface to the *Prolegomena* (1783), "roused him from his dogmatic slumbers,"[5] the solution became possible through a redirection of his enquiries into the area of speculative philosophy.

But it was the analysis of causality, as presented in the wider discussion given to it in Hume's *A Treatise of Human Nature*, that attracted Kant and precipitated his immersal in the problem of knowledge. In a manner typical of the English empirical tradition, Hume had focused on the sensory experience in order to understand the character of knowledge. He concluded that all knowing is derived from the fundamental materials of experience —the heterogeneous conglomeration of perceptions. All that we could demonstrate as being ultimately real was the reception of sensations. The creation of knowledge becomes something other than partaking of ultimate reality. Knowledge is to be found through the operation of such auxiliary psychological principles as constant conjunction, imagination, etc.

Kant was deeply impressed by this deflation of the rationalists' traditional metaphysical claims for necessary knowledge. On the other hand, mathematicians and physicists working with Newtonian ideas were able to make causal implications derived from their theories which were subsequently verified both experimentally and theoretically.

To resolve this paradox Kant eschewed both philosophical extremes—the grasping for a reality in thought beyond the realm of sense experience and the reduction of all knowledge to a scattering of sensory moments explained by superficial and gratuitous regulatory constructs. He recognized the need for the avoidance of the two extreme philosophical positions which could cause us "to oscillate between a scepticism which doubts science because of the failure of metaphysics and a dogmatism which finds in the successful application of our *a priori* categories in science a justification for their application in a region where they cannot be applied successfully. The two evils can be permanently removed only by a 'critical' philosophy, which will show that the categories can be proved, but only for the kinds of objects which we encounter in science and ordinary experience."[6]

Hume had shown through his psychological analysis of the knowledge situation that necessary knowledge was an impossibility outside of the purely analytical areas of mathematics and logic. Instead of meeting this challenge directly, Kant transformed Hume's assertion into a question. How indeed are *a priori* synthetic judgments possible? Implied in this question was a certitude deeply rooted in Kant's commitment to Newtonian science. Cassirer pointed out that this certitude was derived from Euler's *Reflexion sur l'espace et les temps* (in the Berlin Academy of Science's publication of 1748), where Newton's conception of absolute time and absolute space are postulated as physical realities before which all metaphysical and philosophical doubts must be withdrawn.[7] Henceforth, all philosophical knowledge will be based on directives given to us by the mathematical sciences.

This was not a new demand of philosophy. As we have noted, a century earlier Galilean science had similarly stirred the philosophical mind of Europe and had elicited powerful epistemological replies from the writings of Descartes, Spinoza, and Hobbes as well as the later English empiricists. Though Kant was not himself concerned with the historical dimensions of the problem of the status, form, and significance of scientific knowledge, his thinking was conditioned by Hume's presentation of the issues as well as the rationalistic interpretations of Newtonianism. Cassirer postulated that instead of the Newtonian "facts" being interpreted as absolutes, these principles became "conditions of the possibility of experience."[8] Not given to our knowledge like the objects of experience, these principles are forms within which objects are given. These principles in turn lead us to make the judgments through which the necessity of science is established. It was through the development of this structure of knowledge that Kant himself saw nature framed within the Newtonian perspective of universal causality, the conservation of matter, and the continual interaction of all material particles.[9]

When we turn from the problem of the ultimate nature of physical reality to the problem of mind, Kant's special interpretation of idealism comes into focus. He turned radically from Hume's conception of mind as a mere bundle of perceptions and his postulation of certain incidental organizing principles creating knowledge. If there were ever to be any necessity or validity assigned to our knowledge of the physical world, knowledge and mind had to be related conceptually in a more ordered manner. On the other hand, he forswore the rationalistic attempts of Descartes and Leibniz to account for the clear and distinct ideas from which the necessary knowledge of the world of scientific objects was to be built.[10] The great failure of this position lay in its postulation of peculiarly ghostlike sensations that appeared within the epistemological process, having no logical standing because of their rank particularity, yet in some way bridging that irrevocable ontological barrier between mind and matter in the production of these universal concepts.

Kant's solution for these irreconcilable elements—particularity and universality, mind and matter, which only served to thrust the empiricist and rationalist positions apart—represented a veritable Copernican revolution in the theory of knowledge and Kant recognized it as such.[11] Whereas Copernicus ascribed the apparent motion of the heavenly bodies to the motion of the observer on the earth, Kant attributed the apparent characteristics of reality to the active mind of the knower. The problem of the search for the real is transformed into an enquiry concerning the conditions and elements involved in the production of knowledge, knowledge consonant with Newtonian science.

Kant, as did Galileo before him, made the mind an active participant in the creation of knowledge, but did not relegate mind to secondary metaphysical status in the hierarchy of the real. He was willing to confront the activities of thought in the context of human mentation. He did not reduce it metaphorically to a substance, as did Descartes, or to a bundle of perceptions, as did Hume, or even envision it as a passive *tabula rasa*, on which more fundamental metaphysical characteristics (simple ideas) are implanted and set their mark. As Adamson phrased it, "He is not asking, with Locke, whence the details of experience arise; he is not attempting a natural history of the growth of experience in the individual mind; but he is endeavouring to state exhaustively what conditions are necessarily involved in any fact of knowledge, *i.e.* in any synthetic combination of parts of experience by the conscious subject."[12]

1

Establishing the logical conditions for building up scientific knowledge as it takes place within human experience was to Kant but a first step. Next would come examinations of a diverse company of kinds of knowing. Kant's interest in other forms of knowledge derived from a richer cultural frame than heretofore pursued by philosophers. To Cassirer the aim of this endeavor to free knowledge from prior metaphysical and psychological extremes as well as the ultimate fruition of the broader Kantian epistemological enterprise was most clearly expressed in Kant's first *Critique*.

It was necessary to put aside firmly and unequivocally all attempts to penetrate the reality of anything in the domain of knowledge that lay outside experience. This could be done both through an examination of the inner structure of knowledge and by the development through logic of the elements from which thought constructs knowledge.

We must keep in mind the existence of two fundamental and polar elements in every knowledge situation: (a) the given data of sensation, be they experienced as inner or outer sense, constitute the matter of every human experience; (b) we have pure ego or self. The critical teaching of Kant will be seen through an examination of how both scientific and common-sense objects are constructed within the general context of experience.

Kant therefore organized his *Critique of Pure Reason* to deal with this problem by differentiating three main areas—intuition, understanding, and reason.[13] Intuition involves an immediate relation to a given individual object. It is manifested in the passive and sensuous ordering of all empirical perceptions, given in space and time, by our sensibility. This area is taken up in the *Transcendental Aesthetic*. Understanding is the power of thinking, by means of concepts, of the objects given in intuition. The categories of the understanding as they are described in the *Transcendental Analytic* are *a priori* concepts "and depend therefore on the nature of thought, not of things; they are meaningless and empty, except as applied to temporal and spatial things, that is, to appearances given in human intuition."[14]

The area of reason appears in the *Transcendental Dialectic*. Reason, Kant stated, must go beyond what is given to sense. It can conceive of objects which can never be intuited, but these objects can never be established as knowledge. Reason cannot establish truths that go beyond what is given to us in the categories of the understanding, for inevitably it falls into hopeless contradictions, i.e., the Antinomies.

But if these logical aspects are clearly dealt with in discrete sections of the *Critique*, the problem of interpretation is not so easily overcome; it is here that the critical Kant becomes blurred and a variety of interpretations and confusions is literally invited.

Kant was aware that he had arrived at a radically new turn in the knowledge problem. Yet it was difficult to break sharply with the past because of the long linguistic and conceptual history these issues had accrued. Thus his writing contains verbal inconsistencies which mar, if they do not vitiate, the power of his argument.[15] After eleven years of philosophical maturation and writing, Kant quickly organized and set down the *Critique*. It had many structural inconsistencies, not fully rectified even in the second edition, published six years later (1787).[16]

Directed as it is to the problem of the justification of scientific concepts, the entire work focuses in theory on a delimited rendering of what might be called "the epistemological moment." As such, the argument, as it develops from the *Aesthetic* to the *Transcendental Dialectic*, consists of laying out the elements participating in this momentary unity. Not only are these logical polarities established—the form of perception and the form of the understanding—but in the chapter on the schematism, the purely logical analysis of elements is transcended and an attempt is made to account for a sensuous, conceptual unity of phenomena.[17] This ultimate condition under which all knowledge can take place is termed "the synthetic unity of apperception."[18]

Thus, through a logical analysis of experience, we obtain (a) the given data of sense; (b) the forms of sensibility, space and time, within which these data are always experienced; (c) the purely logical and universally applicable categories of the understanding. This finally leads us to a recognition of (d) the ground upon which this synthesis takes place, the unity or identity of *self*, the ultimate condition for all knowledge. Out of this unity we can then extend our perspectives to outer sense, perceived objects, or inner sense, self or ego. The fundamental act of perception and the structuring of experience itself depend upon this " 'transcendental apperception' as a 'condition for the possibility of perception itself.' "[19]

The fact that these elements were developed by Kant as logically separate entities lies at the root of so many of the varied and disparate metaphysical interpretations which Kantianism has nourished. Adamson, among others, strongly emphasized that though these elements are treated as a series of entities they are not merely successive phases of knowledge to be apprehended in their simple succession but are necessarily intertwined as constitutive factors in the process of attaining knowledge.[20]

This difficulty in Kantian exegesis necessitates a special approach to Kant. It is epitomized by Hermann Cohen's comment that it was necessary to do for Kant what Kant himself remarked was necessary for Plato: to "understand him better than he understood himself."[21] There is steady growth in Kant's thinking which is not evident in the body of the *Critique* itself but which necessitates the extraction, from the varying perspectives he here gave the problem, of a defensible and consistent position which would harmonize with his personal development as he studied the problem. It is thus necessary, as commentators from Cohen and Cassirer to Kemp Smith, Ewing, and Weldon have noted, to examine Kant's arguments from the standpoint of this gradual development in his thought.[22] Even though a critical dimension is present in the first *Critique*, it is necessary to go beyond a literal rendering of its message to the historical trend of Kant's thought as well as to the wider corpus of his writings.

2

Very generally we may label Kant's pre-critical phase as realistic. Here the sensations are conceived as the pure given, the solid substratum, which the formative powers of the spirit attack but which they cannot change and the essence of which they cannot penetrate. Sensations, as compared with the purely formal categories of thought, become the matter of knowledge, yet in a different stratum of existence. They therefore can be said to "remain the unpenetrated and impenetrable residue of knowledge."²³ Thus we must postulate a substantial material basis of sensations existing behind the sensible determinants of knowledge. Although the concept of the "thing in itself" was a limiting or negative concept to Kant, Höffding believed that the method and manner in which Kant postulated the ground for our perception of sensations—"things in themselves"—decisively shows Kant to assume an "absolute reality."²⁴ The realism inherent in the postulation of external physical stimuli acting upon our sense organs and brain, which are prior to cognition and possibly to be interpreted as unformed matter (since matter is not yet cognized by the categories in the Concepts of the Understanding), is found not only in the earlier version of the *Critique*, but also in the *Metaphysical Principles of Natural Science* (1786), in which the problem of the constitution of matter itself is raised.²⁵ This generalized realistic tendency in Kant is further epitomized, according to Kemp Smith, in Kant's appreciation of Sommering's theory that "the brain-processes corresponding to the analytic and synthetic activities of the mind consist in the resolution of the water in the brain cavities into its chemical components upon the impact of sense stimuli and their recombination when the stimuli cease."²⁶

As both Weldon and Kemp Smith point out, this realistic tendency in Kant is an inherently unstable one from which it is possible to lapse into a radical subjectivism, depending upon whether one examines that phase of the Kantian "thing in itself" from the ouside—realism—or from the inside—subjectivism. Thus in the *Opus Postumum*, both the world of objects, as they are phenomenologically known, and the phenomenal self, the self-conscious unity underlying this process of knowledge, are called into question. "The phenomenal world thus generated is, to all intents and purposes, viewed as an emanation from the self, creatively produced, but modeled on no previously existent pattern, ideal or real."²⁷ Not only is the subjectivity of space and time accentuated but the forms of sensibility, space and time, are viewed as emanations of mind. "The mind can only anticipate what it has itself predetermined."²⁸

3

There are therefore good reasons for the broad range of interpretations of the Kantian philosophy. But as the Marburg Kantians were wont to reiterate, there is an inner core of Kantian thought which Kant himself labored to unravel and reveal for himself in an explicit and well-defined manner. This can be explained more easily when one starts from the negative.

Cassirer argued that we must declare as illegitimate those explanations in the theory of knowledge which require that the origin of the elements of knowledge be disclosed, or the ultimate ground for their representation. Also, the traditional concern for substantial entities existing as the ultimate constituents of knowledge is to be eschewed. Thus any metaphysical rooting of the Kantian philosophy in Newtonian science and in the logical categories of Aristotle must be understood as not applying to knowledge as it might develop at any and all successive periods in history. This rooting was caused by an unfortunate *a priori* assumption derived from the contemporary historical frame in which Kant lived.[29] Cassirer was at pains to point out the inherent flexibility of the critical position with the development of new scientific conceptual structures, as for example Einstein's general theory of relativity and modern quantum theory.[30]

Instead of asking questions of substance or reality, as for instance regarding space and time, the true critical analysis merely asks the function of the various elements of the knowledge situation.

> But the transcendental philosophy does not have to do permanently with the reality of space or of time, whether these are taken in a metaphysical or in a physical sense, but it investigates the objective *significance* of the two concepts in the total structure of our empirical knowledge. It no longer regards space and time as things, but as "sources of knowledge." It sees in them no independent objects which are somehow present and which we can master by experiment and observation, but "conditions of the possibility of experience," conditions of experiment and observations themselves, which again for their part are not to be viewed as things.[31]

If we accept Kant's view of mind in its critical meaning, i.e., as the active synthesizer of concepts, then the analysis of the process of knowing concerns itself solely with function. Traditional metaphysical expressions of realism, subjectivism, phenomenalism, psychologism, etc., simply disappear.[32] We can argue for this in spite of the unfortunately accurate complaints against Kant's realistic and substantivistic language, as well as his historically rooted blindness in the use of the then current terminology of faculty psychology. The reason for this difficulty apparently lies in the fact that in spite of the radical turn which Kant gave to traditional philosophy he did not completely discard the historical linguistic and conceptual residue. Thus, Cassirer argued, the philosopher deserves to be appreciated for what is best in the sense of his most advanced thought.

The heart of the Kantian first *Critique*, in the critical interpretation given by Cassirer and the neo-Kantians, is, as both Cassirer and Charles Hendel have observed, the section on the "Schematism of the Pure Concepts of the Understanding."[33] It is here that the purely logical deduction of the categories, now in process of schematization, is no longer a "hidden transcendental factor only conceivable in and through analysis." It is a real "phenomenal presence." The problem of the schema resides in the fact that the categories are pure forms of thought. As such they are devoid of any sensuous content. To be ordered into a "unitary consciousness," this sensuous content must be subject to the categories of thought. The problem is to unite the two seemingly unlike factors.[34] This is done by calling into being the transcendental synthesis of the imagination, i.e., the productive imagination. The productive imagination actualizes the categories by providing them with temporal interpretations, since the discrete materiality of the data of sense has, as is shown in the *Aesthetics*, already been given form.[35] The schema is thus a "third thing." It "must be pure, that is, void of all empirical content, and yet at the same time, while it must in one respect be *intellectual*, it must in another be *sensible*."[36] It is the means by which the particularity of the concrete sensuous intuitions are given universality by their subsumption under the category.[37]

As Weldon summarizes the result, "phenomena are nothing more than pure concepts schematized by the transcendental faculty of imagination, and the world of science is a world of schemata, generated in a sense by our own activity but none the less objective in contrast to the 'subjective play of representations' which make up our unreflective consciousness. . . . the categories actualized by the imagination do supply principles which are required for the scientific interpretation of nature and are also essential to the existence of a unitary self-consciousness."[38] It is important even here to reiterate, as did Kemp Smith, that the relationship between the pure concepts, the categories, and the sensible form, i.e., time, in the section of the schematism, should not be interpreted in any other manner than as a "functional relation."[39]

With the postulation of the schema we are catapulted, as it were, into the core of the critical aspect of Kant's epistemology. The schema allows us to redirect our attention to the process of knowledge, whereas the purely logical deduction of the categories presented us with a problem of resolving the differences in the sensuous and formal manifolds. The logical examination in the *Aesthetic* and the first book of the *Analytic* presented us, in the dualism between matter and form, with two seemingly incompatible stems of knowledge, united, Cassirer stated, in an unknown common root or fixed into two irreconcilable states of existence.[40] It is not inconceivable that an invitation was being proferred to a host of metaphysical interpretations concerning the elements of knowledge and their relation as joint concepts of knowledge. But the schematism presents the unity which allows us to concentrate on the process of knowing, allows us to see how out of the object as given to us in the synthetic moment we can find, through logic, a formal as well as a material element. The category of substance could thus be thought of only as a subject and never as a predicate without being schematized, that is, without the temporal and sensible determination of permanence which underlies successive modifications and alterations of accidents. As pure category, it could never represent a natural object.[41] Also the pure category of ground and consequent becomes the schematized category of cause and effect when the transcendental schema of necessary succession in time is applied.[42]

By now focusing upon the objects of cognition, specifically from the perspective of the law of cognition, Kant set the conditions for the constitution of all objects of knowledge. He here firmly established the objective conditions of knowing to be those which are described by the basic concepts of science, namely Newtonian mathematical physics. Thus when we direct our attention to the form of objective knowledge as set down in the section on the schema, we arrive at the established principles of science. We have justified the concepts epistemologically without diverting ourselves into illegitimate realms of metaphysics concerned with the nature of the elements. This critical concern which the Marburg neo-Kantians and Cassirer drew from Kant is with form and function, not with substance and reality.

4

Kant's endeavor to establish the manner in which thought was congruent with scientific knowledge and to demonstrate that a concern for the origins of the ultimate reals of knowledge was not germane to an analysis of the conditions under which deterministic knowledge is formed was not undertaken with the idea of restricting his philosophical perspectives. On the contrary, as Cassirer pointed out, there was throughout Kant's development a parallel concern for what might be described as the humanistic disciplines, a concern which conditioned and shaped his approach to the problem of discursive thought.

Thus on the one hand Kant was trying to eliminate traditional metaphysical concerns in the epistemology of science, at the same time searching for an appropriate context within which to place scientific thought. He was not blind to the fact that the particular concentration on scientific knowledge given certain narrow interpretations of the first *Critique* would lead to an emasculated view of human thought and action. This had been the general trend in the epistemology of the centuries since the birth of science. The trend had oscillated from Galileo's relegation of all sensory, emotive, linguistic, and thus nonmaterial qualities to the nether world of reality and knowledge, to the absorption by Leibniz of all aspects of human thought, science, and the humanities into a theologically rooted and qualitatively structured monadology.

Kant's attempt to reveal the nature and source of objective scientific knowledge was but one step toward an examination of a number of other forms of human knowledge in the light of their own particular laws and orders of construction. In the end, then, there existed an implicit theory of man, but a theory which would be constructed with the same kind of secular analytical procedure that was necessitated in the *Critique of Pure Reason.*

Kant's lectures at Königsberg in the 1760's reflected this early interest and led Cassirer to name him the creator of the discipline of philosophical anthropology. Perhaps the most important influence in this direction came when Kant first encountered the writings of Rousseau. Kant's *Enquiry into the Evidences of the Principles of Natural Theology and Morals* (1763) and the *Observations on the Feeling of the Beautiful and the Sublime* (1764) attest to the immediate influence of Rousseau and the manner in which the latter's writings helped Kant to see more clearly his own developing conception of man.

Kant's sensitivity to the various philosophical, literary, and artistic movements of his rapidly changing times allowed him the perspective to keep these various stands logically distinct, to keep the rationalistic form-centered attitudes from being absorbed into the growing feeling-oriented concerns represented by the *Sturm und Drang* of the German poets.[43] In the presence of a growing romanticism in Germany, Kant's concern with the exploration of the formal element of cultural expression constituted a conserving rationalistic element, in that he seemed to restrain intellectually some of the more extreme tendencies. His intellectual theories are therefore hardly the invitation to a volitional personalism or a philosophical antirationalism that has traditionally been imputed to him.

The theme that comes through in all his writings and that is in tune with the critical philosophy as expressed in the first *Critique* is his resistance to the reductive seductions of metaphysics and his search for the central core of meaning, the ideal truths, out of which each realm of human experience is built. "And in Kant's system an ideal is not, as with Plato, something opposed to experience—something lying outside it and elevated above it. It is rather a moment, a factor in the process of experience itself. It has no independent, isolated ontological existence; it is a regulative principle that is necessary for the use of experience itself, completing it and giving it a systematic unity."[44]

If there is a central theme in Kant, Cassirer felt it was his assertion of the autonomy of human nature and knowledge. Kant's views exemplify an intellectual resistance to the various attempts to reduce man to some externally necessary principle, attempts which had plagued all of philosophy since the scientific revolution. Perhaps this resistance was the reason for Kant's deep interest in Rousseau.[45] For in Rousseau's search for natural man, *l'homme de la nature*, there is no imputation of a historical or biological rooting of human nature. He is attempting to free the conception of man from the limitations of contemporary usage, *hommes des hommes*. It is for conceptions such as these that Kant called Rousseau the Newton of human nature. Rousseau wanted to educate man, as in *Emile*, outside society, in nature, so that he might enter a truer society, wherein exists a vision of law, of right, of freedom.[46] But to examine man in his deepest involvements with creativity, freedom, and autonomy, an analysis of the intellectual problems of the day is necessary.

Kant believed the philosopher must show how man transcends the mechanical vision of living processes, which was even then so powerful in its rational persuasions. Man's assigned station must be conceived as being located not in nature alone, "for he must raise himself above it, above all merely vegetative or animal life. But it is just as far from lying somewhere outside nature, in something absolutely other-worldly or transcendent. Man should seek the real law of his being and his conduct neither below nor above himself; he should derive it from himself, and should fashion himself in accordance with the determination of his own free will. For this he requires life in society as well as an inner freedom from social standards and an independent judgment of conventional social values."[47]

Cassirer noted that this view of Kant's was the product of Rousseau's inspiration on human nature, society, and education. But while it allowed Kant to turn from the imperious demands of the physical sciences to wider areas of human behavior, it did not propel him into postulating complete and chaotic freedom in ethics, esthetics, or religion. Discursive and objective thought was no arbitrary accident of human nature, merely the exemplification of the law and form-giving activities of the mind in one among a number of different areas of cultural thought. Kant's postulation of laws of ethical behavior is not a matter of imposing normative "oughts" upon man because he personally felt that these laws were good for man. The search for the laws of human behavior is a manifestation of Kant's conviction that a unitary vision of human nature can accommodate itself to the variety of man's concerns and activities. Therefore, there must be principles in each of the various areas of knowledge.

Thus while Rousseau provided the inspiration for the program, Cassirer pointed out that Kant needed to subject this inspiration to careful philosophical analysis.

He demanded definiteness and accuracy in ideas and clarity and perspicuity in their architectonic construction. He had to think Rousseau's ideas further, and he had to complete them and give them a systematic foundation. And in so doing it developed that this foundation led to a problem of absolutely universal significance, to a problem that included a genuine "revolution in men's way of thinking." Only through a critique of the entire "faculty of reason" could Kant solve the conflict that had inspired Rousseau in his fight against the *philosophes;* only in this way could he create that wider and deeper idea of "reason" which could do justice to Rousseau's ideas and incorporate them in itself.[48]

This, then, is the critical Kant to whom the Marburg Kantians, and especially Cassirer, looked as the inspirer in their search for an understanding of the various dimensions of human thought. The pre-Kantian philosophic involvement with man as knower exhausted only one aspect of man's functioning. Kant believed the human spirit to be capable of creating other forms of objectivity in addition to science. In examining man's ethical, artistic, and organic dimensions in the *Critique of Practical Reason* and the *Critique of Judgment*, Kant attempted to broaden and organize our understanding of the various creative perspectives that exist in man. Cassirer described this Kantian ideal as follows:

> This *gradual* unfolding of the critical-idealistic concept of reality and the critical-idealistic concept of the spirit is among the most characteristic traits of Kantian thinking, and is indeed grounded in a kind of law of style that governed this thinking. He does not set out to designate the authentic, concrete totality of the spirit in a simple initial formula, to deliver it ready-made, as it were; on the contrary, it develops and finds itself only in the progressive course of his critical analysis. We can designate and define the scope of the human spirit only by pursuing this analytical process. It lies in the nature of this process that its beginning and end are not only separate from each other, but must apparently conflict—however, the tension is none other than that between potency and act, between the mere "potentiality" of a concept and its full development and effect. From the standpoint of this latter, the Copernican revolution with which Kant began, takes on a new and amplified meaning. It refers no longer solely to the function of logical judgment but extends with equal justification and right to every trend and every principle by which the human spirit gives form to reality.[49]

This demand that the dimensions of human knowledge be extended to the various spontaneous formative capacities and interests of man can be applied equally well to the epistemological quest Cassirer set for himself in *The Philosophy of Symbolic Forms*.

3

Newtonianism and the Status
of Scientific Knowledge

KANT'S ABILITY TO APPLY HIMSELF TO A DIVERSE
series of cognitive problems in an acute and logically insightful
way is reflected in his great impact on the intellectual world.
There are many Kants, depending on the particular interests and
inclinations of the interpreter. There was in his own day the Kant
of Schiller, of Goethe, even of Beethoven and of course of
Hegel.[1] In our own time we have had innumerable interpreters
and schools which have projected back upon the philosopher
their own particular interpretive colorations of the important
principles Kant labored so arduously to bring forth.

Kant's three *Critique*s are significant for their unity of perspective. The task of imparting his message was a difficult one, partly
because of the special revolutionary character of his philosophy
and partly because of Kant's use of the language of outworn
metaphysics and faculty psychology, which obscured his purpose
from many followers. The distinction he made between ontology
and logic and the special regulative status of metaphysical concepts is difficult to accept for those bent on contact with ultimate
reality. His anthropology, which saw in human nature mental
correlates of the three logical forms of thought—understanding,
reason, and judgment (cognition, desire, pleasure)—though they

are derived from presumably empirical sources, was limited and timebound in that it reflected contemporary usage.

There was, finally, in the post-Kantian intellectual environment, a falling away from the delicate tension between the unity of vision in the critical philosophy which bound the *Critiques* into a panoramic perspective of human thought and the separateness which was implied in the various faculties of thought, each with a special logic of its own, yet in no way separated in the empirical sense. This was an era in science in which discrete areas of physical experience were being explored in order to build up a structure of interrelated laws through the analysis of these special areas. On the other hand, a poetic and indeed metaphysical vision which by the end of the eighteenth century was taking hold of the European mind was intent on absorbing the vast diversity of things into an all-encompassing intuition of reality.

One might make a parallel here between the balance that Kant achieved in logic and epistemology and that delicate if tenuous relation between reason and faith achieved much earlier by St. Thomas, which itself was subject to the vicissitudes of time and criticism.[2] The Romantic of the late eighteenth century listened with his inner ear to Lessing, Herder, and then to Schiller and Goethe, while paying intellectual obeissance to Kant. First, under Fichte and Jacobi, the concern for human freedom which Kant had presumably rescued from the mechanists in the second *Critique* became the central object of philosophical thought. In the fruition of Schelling and Hegel the preoccupation with the "thing in itself—*noumena*" led to a vision of the primacy of mind, now conceived metaphysically as spirit, encompassing all of nature, man, and history. The regulative function of metaphysical ideas was transmuted into constitutive characteristics of things as they really were.

Kantian philosophy thus stood at the culmination of the Enlightenment and, if it did not undergird, at least prepared the way for the two divergent streams of thought which dominated the nineteenth century—physical sciences (Newtonianism) and academic philosophy (Hegelianism). The Kantian synthesis was dis-

solved and each movement pursued the special quality of intellectual concern appropriate to the logical techniques available.
Of the two, Hegelianism is the least amenable to the traditional
critical Kantian attitudes, although it should be pointed out that
there is much to remind one of Hegel's philosophical enterprise
in Cassirer's later synthetic work, *The Philosophy of Symbolic
Forms*. It is Hegel's intuitionism which departs most from the critical method, in that there is never the test of experience and fact
to give the interesting esthetic qualities of his philosophy relevance in the empirical realm.

By limiting their range of concerns to physical experience and
disciplining their methodology through mathematics and experiment, the scientists created a body of knowledge that hopefully
could be indefinitely enlarged and contributed to in many diverse
ways and yet not be bogged down in metaphysical disputation. Symbolic of this separation of the two streams of intellectual endeavor was the founding of the Ecole Polytechnique in
1794–95, which in effect established science as a professional
endeavor, although outside the traditional university regimen.
Meanwhile, in Germany, academic philosophy dominated the
universities.[3] For philosophy this state of affairs was perhaps permanently debilitating, for the valuable Hegelian intellectual
esthetic was abandoned along with the execrable Hegelian rhe-
toric. The physical sciences, however, continued to grow richer
in cognitive beauty as well as in empirical and practical import.

At the time Kant was writing the first *Critique* the Newtonian
doctrine was being gradually absorbed into the fabric of European intellectual life. By the latter part of the eighteenth and the
early part of the nineteenth century, the pace of scientific discovery itself speeded up considerably. New realms of scientific
research were opened with the work in chemistry of Lavoisier
and Dalton and in electricity of Galvani and Volta.

The contrast of these realms of experience with the traditional
Newtonian astronomical context represented to the scientific
mind a challenge to expand the scope of Newtonianism. It was
now believed that these laws could encompass all physical phe-

nomena in our universe. The small size of the component masses and distances thus represented a stimulus and an ideal for Coulomb, who in 1788 postulated a set of equations after the Newtonian model to represent electrostatic action.[4] His was one of the earliest manifestations of Laplace's ideal that all motion be determined by one law's application.[5]

Probably the most important advance of the Newtonian mechanical model to the status of a philosophical principle occurred in Ampere's electrodynamic equations (1829), which to an amazingly successful degree applied the mathematics of Newtonian mechanics to electrical motion.[6] The power of this intellectual ideal radiated over the entire scientific world. Theories which opposed certain Newtonian conceptions, now revived by Ampere's experiments, were downgraded—such as Huyghens's wave theory of light, which contrasted with Newton's corpuscular conception. The kinetic theory of gas originally set forth by Daniel Bernoulli in his *Hydronamica* was likewise displaced by the atomism of Berthollet and Dalton, only to be revived in about 1850 by Clausius and Joule after a lapse of over one hundred years.[7] Not before the mid-nineteenth century work of Joule were the mechanically biased substantialistic theories of heat rejected. Their eventual supplanter, the kinetic theory of heat first set forth by Count Rumford in 1798, was then virtually ignored, so strong were the Newtonian influences.

In this way the principles of mechanics and the concepts of force and mass began to be hypostatized into the "reals" of the universe. They were endowed with absolute truth and to the scientific mind became the key to understanding the nature of reality. With this turn toward a metaphysics of absolute realism, the older epistemological problems inherent in the historical tradition of Galileo, Descartes, Locke, and Leibniz now became centered in the physical sciences.[8] The lesson of Kant's critical philosophy—especially its emphasis on the need for restraint in judging the status of scientific principles—was lost in the rush of new empirical materials.

To Cassirer the resulting philosophical controversies constituted the second important phase in the development of the theory of knowledge. Not only would the context of debate over physical theory develop a resurgence of metaphysical thinking—admittedly localized in a delimited area of scientific experience—but by the end of the century the truths inherent in Kantianism would again be expressed, both in the context of a self-consciously Kantian school of philosophy and by a number of other philosophers of science coming to the "critical position" independent of any historical awareness of its sources.

1

Newtonianism from the time of D'Alembert (c. 1750) maintained a strong hold over the scientific community until well into the middle of the nineteenth century. Its persuasive power lay in a simple if absolute realism which seemed to satisfy the philosophic element of thought and in the enormous predictive power of mathematics in a variety of experiential contexts. In contrast to prior philosophic systems, content with explanation alone, the future intellectual status of this world view depended upon the extent to which it fulfilled its experimental promises.

It was exactly as might be expected. The deficiencies of mechanics in explaining and ordering phenomena would in the end undermine the philosophical assumptions of those who earlier had professed adherence. These explanatory difficulties can be seen as far back as the work of Young, Arago, and Fresnel, who argued against Newton for Huyghens's wave interpretations of light. In spite of their postulation of the ether as transporter of these waves—a new nemesis for scientific progress—their views constituted an early and pregnant deviation. Parallel to this, Oersted and Faraday, by locating the lines of force within the field, also with the help of the ether concept, placed in opposition to the then dominant substantialism of electrical fluids and the idea of action at a distance a theory of continuity which led eventually to Einstein's work in displacing mechanics from its absolutistic position.[9]

However, the tenacious philosophical power of mechanical theory and its Newtonian ideal was a powerful inhibitor of new theories. This can be seen in the fact that though Maxwell, in writing his revolutionary treatises on electromagnetism and light (1860–1870), rendered irrelevant the traditional corpuscular constructs essential to the visualization of mechanical theory, the models he utilized for explanatory purposes were still mechanical. Faced with this dilemma of interpretation Hertz was forced to state that the meaning of Maxwell's theory lay in Maxwell's equations alone.[10] A man as philosophically sophisticated as Max Planck could still hold, in 1887, that the principle of the conservation of energy could be derived from mechanics. He could no longer make such claims in 1910 after the first papers on relativity were published.[11]

However, as Cassirer constantly pointed out, as the nineteenth century continued, the list of scientific detractors from Newtonian absolutism grew steadily. The defenders of the mechanical perspective of reality, who numbered among their group such figures as Wilhelm Weber, Helmholtz, Lord Kelvin, E. du Bois-Reymond, by the end of this period found themselves expounding virtually a minority point of view. Scepticism of the absolutes of space, mass, and force was finally epitomized in the incisive attacks by Mach and Ostwald on the supposed reality of the atom.[12] Dumas had questioned this realistic interpretation of the atom as early as the 1840's and concluded that science should not transgress the limits set by experience. In fact the gradual superimposition of the various conservation theories upon the independent substantiality of the many fluids—caloric, electric, magnetic, etc.—was accomplished because of a growing if unexpressed suspicion of the supposedly "independent reality" of these conceptual entities.[13]

A high point in the growing sophistication of scientists regarding the status of their theories is represented by Kirchhoff's denial that the mechanistic view provided an explanation of natural phenomena. The goal of mechanics was merely to give, in the simplest possible fashion, a complete description of the motions occuring in nature.[14] In this he was close to Boltzmann's later denial of the absolute existence of atoms, which Boltzmann claimed were not to be conceived as more than "images for the exact representation of phenomena,"[15] or Boltzmann's approval of Kirchhoff's view that theories "bear the same relationship to Nature as symbols do to the entities they stand for," a pregnant conception taken up later by Duhem.[16]

There was, however, nothing intrinsic to mechanical theory alone to attract so many of the metaphysical extrapolations here mentioned. Cassirer noted that the science of energetics, which can be traced back to the kinetic mechanics of Huyghens and Leibniz but made its greatest advances in nineteenth-century thermodynamic theory, was also subject to this kind of philosophical debate.[17] Thermodynamic theory is characterized by its resistance to any attempts to visualize underlying entities. The peculiar mathematics involved allowed this neglect of representation, because the "equations of the theories expressed direct connection among the magnitudes observed."[18] In spite of this, a scientist such as Ostwald, while attacking the reality of the atom (by labeling it a metaphysical construct), could turn about to reify energy itself as a substance. The positivist reaction to this view given by Helm, also a believer in the primacy of energetics, represented the other extreme in metaphysics to which scientists were prone. "There exist neither atoms, nor energy, nor any kind of similar concept, but only those experiences immediately derived from direct observation."[19] On the other hand, both Rankine and Mayer, in resisting all attempts to attribute qualitative characteristics to this science (thermodynamics), could express a moderate epistemological position. "Precise delimitation of the natural limits of human inquiry is an end that has practical value for science, whereas the attempt to penetrate by hypotheses to the inner recesses of the world order is of a piece with the efforts of the alchemists."[20] Thus, faced by the extremes of philosophical reductionism, partisans of a variety of physical systems would in the end return to a critical and methodological approach to science.

2

These concerns with the nature of the physical sciences that plagued members of the profession eventually led to an interested response in the philosophers. Edward Zeller, writing *On the Significance and Problem of the Theory of Knowledge*, which appeared in 1862, not only coined a new term (theory of knowledge) but signalled a philosophical revival of Kantianism as part of an attempt to clarify issues in science which had stirred up so much debate.

As a philosophical tool the theory of knowledge would be used to sweep away the traditional Hegelian pretensions of "reaching to the heart of absolute being." Zeller denied the Hegelian claims that all knowledge could be encompassed in one vast metaphysical gulp "because it overlooks the conditions of human knowledge, for it purported to grasp with one swoop from above the ideal of knowledge which, in reality, we can approach only gradually through complicated labour from below."[21] It was the task of logic to help us clarify the problem of knowledge of reality "by analyzing and tracing to their original conditions the ideas of the human mind which form the basic material for all empirical knowledge."[22] Henceforth, as Cassirer described, the efforts of the neo-Kantian tradition would be to lead philosophy, in Kant's words, "into the safe road of a science" to make the process of the formation of scientific concepts the central problem, rather than the pursuit of the inner essence of reality.

Of the scientists intellectually equipped to deal philosophically with their discipline, H. L. F. Von Helmholtz was perhaps the most universal genius of his time. It was Helmholtz, Cassirer tells us, who during this mid-nineteenth-century period led the "back to Kant" movement. Not only was he an important leader in physics but he was also renowned in the area of the psychology and physiology of sensation. As a philosophical interpreter of his scientific disciplines he represented a halfway point in the gradual articulation of the Kantian critical position on knowledge for the new physical sciences.

In the spirit of Kant, Helmholtz affirmed that a theoretical understanding of the significance of what was being achieved empirically in the laboratory was always of first importance and that the "establishment of the presuppositions of science is a common task of physics and philosophy that no era can escape with impunity."[23] The catholicity of his interests, as contributor to the study of physiological optics as well as to mechanics, enabled Helmholtz to appreciate the active and constructive role of the human mind in building the various forms of knowledge. Thus he opposed the theories on sensation and color of both Hering and Katz, for whom the perceptions of various types and modes of color phenomena were rooted in the inert concepts of memory or reproductive imagination. Following the Kantian theme, Helmholtz saw the mind as an active collaborator in the reception and ordering of sensory experience. He postulated that the various supposedly intuitive images of colors, shadows, textures, and distances, were products of judgments and inferences and were, therefore, logical and intellectual rather than passive or reproductive in function.[24]

However, as Cassirer pointed out, in this earlier phase of neo-Kantianism, both Helmholtz and Zeller were under the influence of what might be called "the deficiencies of the empirico-physiological interpretation of Kantian a priorism."[25] Thus, to Helmholtz, Müller's discovery of specific nerve energies was "the empirical realization of Kant's theoretical explanation of the nature of human understanding."[26] Again, Helmholtz believed that "the true and permanent achievement of Kantian philosophy . . . consists in having shown the participation of the innate laws of the mind in the formation of our ideas."[27]

Zeller phrased the aims of the neo-Kantians in somewhat the same manner. "The way we have to proceed to obtain exact concepts, can be decided by reference only to the conditions under which the formation of our concepts takes place according to the nature of the mind; it is just these conditions that the theory of knowledge ought to investigate and to determine, accordingly, whether and with what presuppositions the human mind is capable of a knowledge of truth."[28]

But if Helmholtz's wide-ranging empirical and intellectual concerns allowed him to perceive the importance of Kant and his theory-making for scientific endeavors, it was too early in the development of both science and its new philosophical interpretations for him to achieve a full and consistent integration of Kantian principles. We can see this in Helmholtz's theoretical articulation. All knowledge consists of signs which are brought to our awareness through experience. Our concepts of the world are formed out of these signs. For Helmholtz, "the world of phenomena is nothing other than an aggregate of signs, which are in no wise similar to their causes, real things, but which are lawfully correlated with them in such a way that they can express all differences and relations of things."[29]

However when dealing with the phenomena of sensations Helmholtz was constantly pulled by the subject matter at hand into a quasi-Berkeleyan empiricism. Originating in the sense perceptions we receive, knowledge is formed, he stated, because these "sense impressions are signs for our consciousness, the interpretation of which is left to our intellect."[30] It is true that the world of phenomena becomes "nothing other than an aggregate of signs" and "what lies within our reach is knowledge of the lawful order in the realm of the real, and this to be sure only as represented in the sign system of our sense impressions."[31] Thus, as Cassirer pointed out: "In this view the logical concept accomplishes nothing other than to fixate the lawful order that already lies in the phenomena themselves."[32] And finally, as Helmholtz himself stated it, the representation of an object in a concept "embraces all the possible aggregates of sensation which this object can call forth when regarded, touched, or otherwise examined from different sides."[33]

There is a reality beyond our knowledge, Helmholtz posited, but we can never penetrate its inwardness or see things as they are objectively, since our knowledge consists only of signs satisfying certain conceptual conditions and demands.[34] Phenomena as signs are not "similar to their causes, real things," Cassirer pointed out about Helmholtz's avoidance of any correspondence theory, but "are lawfully correlated with them in such a way that they can express all differences and relations of things."[35] The inferences that man makes in interpreting these signs reveal the Berkeleyan empiricism latent in Helmholtz's thinking, since they derive from inductive patterns of inference where associative combination and reproductive completion of sense patterns suffice to explain the articulation of these signs in the various orders of perception.[36] The important difference between Berkeley and Helmholtz is that in Helmholtz there is an ontological gap between the order of pure sensations and the order of signs. Unlike Berkeley, the sense impressions themselves can never take on a representative function beyond their presentative content.[37]

If Helmholtz retains a measure of the Kantian feeling in his *Optik*, his more general commitment to Newton, which makes him the most devout of the mechanists, brings him more fully into the metaphysics of materialistic realism. The ultimate reality is matter. "Matter becomes the fundamental reality which has to be described in our scientific concepts of nature."[38] Matter is here conceived in Newtonian terms. Thus in Helmholtz's classic, *On the Conservation of Energy*, all happenings in nature are reduced to attractive and repellent forces. This proposition is not stated as "mere *fact*" but its validity and necessity follow from "the *form* of our understanding of nature."[39]

This important law (conservation of energy) is seen as completely subsumable to and derivable from the theories of mechanics, not vice versa, as many energeticists were soon to claim. In addition, as Cassirer pointed out, "instead of subordinating the problem of causality to the general problem of signification," as a true Kantian would, "he takes the opposite course, interpreting the function of the sign itself as a special form of causal relation and seeking to explain it as such."[40] The recognition of the category of causality now became a logically necessary precondition for the subsequent intelligibility of nature for man. In these attitudes of Helmholtz towards physics and mathematics, so Cassirer postulated, we see an attempt to balance Berkeleyan empiricism, the Kantian theory of signs, and a traditionalist Cartesian intellectualism.[41]

What Helmholtz had accomplished in his theory of knowledge was to acknowledge a reality existing apart from our only source of knowledge—signs. However, for mechanics to give us knowledge of the world it was necessary that we ascribe overriding philosophical preeminence to the reality of Newtonian science. Even though his conviction as to the nature of the content we call knowledge was clear, i.e., that knowledge is comprised of intellectually created signs, we must once more make this human activity subservient to the larger, external designs of the universe, even though our access to an intuition of the real world must come forever through signs. In some way man's knowledge is still subservient to the reality that confronts him with this knowledge. The logical implications of this metaphysical paradox were not long in coming, even from those still adhering to the mechanistic ideal of scientific knowledge.

As early as 1872, Emil du Bois-Reymond, in his "Über die Grenzen des Naturerkennens," spoke of the ultimate epistemological limitations of scientific theory in which reality must remain forever inaccessible to scientific knowledge.[42] He was willing, Cassirer noted, if Helmholtz was not, to accept the implications of a neo-Kantian critical position on the nature of all knowledge, including mechanics. If scientific knowledge is mechanical knowledge, then the limitations of mechanical knowledge in not being able to encompass all phenomena must further illumine for us the existence of a realm in which science cannot probe. He would accept limitations to mechanics presumably because he now accepts the metaphysical limits about what one would claim for mechanical theory. However, in spite of this ultimate barrier du Bois-Reymond declared, "For us there exists nothing but mechanical knowledge, no matter how miserable a substitute it is for true knowledge, and accordingly only one true form of scientific thought, that of mathematical physics."[43]

His brother, the mathematician Paul du Bois-Reymond, drew an even more decisive Platonic picture in *Über die Grundlagen der Erkenntnis in den exakten Wissenschaften,* in which physics is doomed to failure from the beginning in its attempt to grasp reality.[44] Each attempt teaches us "how impenetrable are the walls of our intraphenomenal prison . . . . Our thinking, which wears itself out attempting to advance in an area of misty uniformity, is as if lamed and does not take a single step forward. We are encased within the housing of our perceptions and are as it were born blind for what lies outside. We cannot even get a glimmer from the outside, for a glimmer would already be like light, and what in the real corresponds to light?"[45]

By the beginning of the twentieth century even the limited claims for mechanics and materialism had been undermined and the admission of an ultimate "ignorabimus" had ended in an outright plea of desperation with Abel Rey. "The failure of the traditional mechanistic science . . . entails the proposition, 'Science itself has failed.' We can have a collection of empirical recipes, we can even systematize them for the convenience of memorizing them, but we have no cognition of the phenomena to which this system of these recipes are applied."[46]

Pierre Duhem, acting in the self-appointed role of a father confessor to the writer of this famous doctoral thesis, compassionately agreed with Rey that critical reflection had lowered theoretical physics to the status of knowing only the experimentally revealed truths, bound to contingency and particularity. However, in spite of this, the scientist cannot accept the view that in physical theory we have "merely a set of practical procedures and a rack filled with tools." Physical theory does confer on us a "certain knowledge of the external world which is irreducible to merely empirical knowledge." Duhem postulated "natural classification," which he felt had a status beyond practical and aesthetic values and which explained its power over the mind of the scientist.[47]

3

The disillusionment with the philosophical claims for mechanical theory was an important step in freeing science from the more gross realistic and metaphysical assertions that had been made in the first and subsequent bursts of enthusiasm over the theory's usefulness in guiding and synthesizing the various experimental domains. Cassirer hypothesized that of all the great scientists of the late nineteenth century Ernst Mach was the most influential in puncturing these claims for the absolute status of Newtonian theory. He was aided not only by the disintegrative tendencies in mechanics, e.g., the notable failure to discover the ether experimentally, but by the great growth of field theories, which remained unaccounted for by classical mechanics.

Mach directed his attack on mechanism along a wide front. The primary object of his scorn was the concept of the atom, and here, along with Ostwald and other energeticists (e.g., Rankine), he stressed the speculative character of this unobservable entity. Indeed the very idea of matter or of materialism, which was the object of great debate among the philosophers at that time, as documented in Lange's *History of Materialism*, was anathema to Mach. Likewise he denied the reality of an electrical substance, a concept much touted at that time by Sir Oliver Lodge.[48] The entire system of Newtonian absolutes, about which so many doubts had begun to accumulate, was questioned. Indeed, Mach's criticism of the concepts of force and mass, that they were defined in such a way as to "leave us in a logical circle," as well as the reconsideration of Newton's bucket experiment, showed "how close to his way of thinking was the search for relativity in a general sense," (relativity of acceleration).[49] Mach's view of the law of gravitation as nothing more than a summation of a large number of concrete observations given a common linguistic expression was typical of the manner in which he hoped to avoid any semblance of a realistic interpretation of physical theory.[50]

Physical theory, Mach believed, begins with experience and experience expresses the structure of the world. What is the structure of the world that experience gives us? "The world consists of colors, sounds, temperatures, pressures, spaces, times,

and so forth, which now we shall not call sensations, nor phenomena, because in either term an arbitrary one sided theory is
embodied, but simply *elements*. The fixing of the flux of these
elements, whether mediately or immediately, is the real object
of physical research."[51] Within the flux of experience we note
certain regularities. These regularities are not accidental but are
universal aspects of our thought. Physical theory is a description,
not an explanation of these regularities. In Mach, knowledge is
built up through a process reminiscent of Galileo's compositive
method. Through analysis and synthesis, separation and recom-
position, the theory takes form. Thus "all general physical concepts and laws, the concept of the ray, the laws of refraction,
Mariotte's law, etc., etc. are obtained through idealization. In
this way they take on that simplified form . . . which makes it
possible to reconstruct and therefore understand any fact even
of a complex nature by means of a synthetic combination of these
concepts and laws."[52]

According to Cassirer, the overriding theme of Mach's philosophy of science is that the ultimate facts are the sensations we
find all around us. In contrast to Helmholtz, Mach stated that
"sensations are not signs of things; but on the contrary, a thing is
a thought symbol for a compound sensation of relative fixedness.
Properly speaking the world is not composed of 'things' as its
elements, but of colors, tones, pressures, spaces, times, in short
what we ordinarily call individual sensations." The object of
science, in Mach's view, is "to replace, or save experiences, by
the reproduction and anticipation of facts in thought. . . . This
economical office of science, which fills its whole life, is apparent
at first glance; and with its full recognition all mysticism in science disappears."[53]

On occasion Mach hesitated, stating that we reduce all the
elements of experience, as they are referred to our own conscious experience, to sensations. But as these elements refer to
external "objects" (the materiality of physics), we may state that
they are elements or neutral entities. That all physical theories
must be erected from these observable elements is a universal

demand of this phenomenalistic physics. It leads to interesting characterizations of some of the least complex abstractions, such as "body."

> A body looks differently in each illumination, gives a different optical image in each position, gives a different tactual image with each temperature, etc. However, all these sensuous elements so hang together, that with the same position, illumination and temperature, the same images recur. It is thus a permanency of the *connection* of sensuous elements, that is involved here. If we could measure all the sensuous elements, then we would say that the body subsists in the realization of certain equations, that hold between the sensuous elements. Even where we cannot measure, the expression may be retained symbolically. These equations or relations are thus the really permanent.[54]

The Humean quality of the above statement is not difficult to identify. Cassirer noted that, in a sense, it is a return to the nominalism of the empiricistic tradition, especially in Mach's treatment of such things as causality and substance.

The reaction to Mach's program was mixed. While many philosophers were enthusiastic about his positivistic criteria and felt that the development of modern scientific empiricism could be traced to Mach, others, including the neo-Kantian philosophers in Germany, saw his views as an epistemological regression not only to the English empiricism of Mill but back to Hume and Berkeley. The scheme to unite psychology and physics—as contrasted to Helmholtz's dichotomization of these areas—through the postulation of the so-called neutral entities (in reality sensations viewed from differing perspectives) was thought by his critics to be of dubious historical value.

The first basic problem in Mach's program relates to the reality or ultimate facticity of sensations. Philipp Frank maintains in several papers that Mach never meant to substitute for the hypostatizations of the materialists any ultimate description of reality (sensations) but only to indicate the need for connecting all physical theories by minimal theoretical links to sensory experience and experimentation.[55] By doing this, Mach, so Frank argues, would eliminate all metaphysical constructs such as matter, life, soul, collective soul.[56] Although, as Ernest Nagel says, it would involve rather unimportant matters of textual exegesis, there are many quotations which could be cited which explicitly connote a realistic designation of ultimate facticity and reality to sensations.[57]

Cassirer, who became a member of the Marburg school during the height of this debate and who was thus concerned with the challenge to the neo-Kantian program, appraised Mach's sensationalism "as a substance, hypostatized as the universal matter of the world."[58] Mach brought back into the heart of physics a new form of metaphysics. Cassirer often paraphrased Goethe in expressing his opposition to the positivistic search for ultimate facticity: "there is no fact without the problem of formation: all fact is in itself theory."

Mach's position likewise was not supported, as Frank notes, in his emphasis on the nature of hypotheses and the necessity of postulating hypotheses which are linked directly to phenomena that can be perceived. D'Abro, for example, discussed at length the manner in which use of microscopic unobservable entities has been of great value in science, from Bernoulli's kinetic theory of gas and statistical characterization of the law of entropy to Lorentz in his supplementation of Maxwell's equation.[59] Max Planck's position was that Mach's views on this matter, instead of purifying and reinvigorating theory, led to the "dissolution of knowledge."[60] The general opinion was that any physical theory which bore fruit for further investigation must of necessity abstract from perceptual experience, if only to free the mind for new insights.

Boltzmann joined Planck in criticizing the restrictions on the hypothetical constructs of science, declaring that on the basis of Mach's views the concept of the star was just as real or unreal as the atom.[61] In fact the discovery of a series of empirical confirmations of the construct of the atom, such as Boltzmann's reestablishment of the kinetic theory of gases, the investigations of

Brownian movement, Helmholtz's proof of the atomic nature of electricity, and Laue's discovery of crystal diffraction, finally induced even Ostwald, a vigorous opponent of all atomism—yet who had created what amounted to his own metaphysical version of substance—energy—to declare "experimental proof of the atomic structure of matter had at last been attained."[62]

Mach's phenomenalism was not completely unappreciated. The logical positivist movement, as Frank points out, found much in Mach to its liking.[63] And the thunderbolts they unleashed against metaphysics in support of a view of philosophy synonymous with the logical canons of physical science were inspired to a great extent by Mach and the entire empiricist movement of the late nineteenth and early twentieth centuries. Indeed, Rudolf Carnap, who could perhaps be considered the doyen of the movement in its "classical Viennese" formulation, resorted to a sensationalist epistemological criterion for physical theory, in his *Logische Aufbau der Welt*.[64]

Thus "the meaning of a statement in science would be the sum of all statements about similarity and diversity between sense impressions that can be derived logically from the statement in question."[65] Later Carnap rejected this view, perhaps again illustrating the dangers inherent in any doctrine which claims to have eliminated from philosophy all vestiges of metaphysical statements, in that metaphysics can be a double-headed axe and can fall in more than one direction.[66] Thus, in its own peculiar manner, sensationalism too is a derivative, "verbal" formulation of experience, far removed from the supposedly direct and immediate contact with the given of experience.

Frank further articulates the importance of Mach's work for Einstein and the creation of the theory of relativity through Mach's insightful criticism of the Newtonian absolutes, mentioned previously. Einstein, however, explicitly rejected Mach's linkage of theory with experiment or observables, saying only that somewhere in the freely constructed theory there should be a link with direct experience and that it should be capable of being tested experimentally through the medium of certain consequences.[67] In this view even the so-called operationism of Percy Bridgman, with roots in the Machian point of view, is rendered restrictive of the free symbolic construction of physical theory, for in practice it is not and has never been possible to test all the assertions of any theory.[68]

In general, critics such as Cassirer are inclined to accuse Mach of disregarding his own strictly positivistic and phenomenalistic canons. Perhaps the bitterest remarks are contained in a work of E. Study, who derides Mach's supposed philosophical utopianism and states that such a program is impossible to apply consistently and has never been and never will be realized.[69] Not only is Mach's program impossible, but he invites in at the back door what he locks out at the front.

> In numerous cases the hypotheses that are basely denounced at the official reception (why not atomists, too?) are admitted, under a different name and through a back door especially arranged for this, into the sanctuary of science. Such names and corresponding motivations are by no means few. Without any effort the writer collected a full dozen of them: "most complete and simple description" (Kirchhoff), "subjective means of research," "requirements of conceivability of facts," "restriction of possibilities," "restriction of expectation," "results of analytic investigation," "economy of thought," "biological advantage" (all of these employed by E. Mach).[70]

Perhaps Mach's greatest failure, according to Cassirer, in addition to his metaphysical reification of sensations and the failure of his nominalism to advance the progress of physical theory, was his departure from his own canons. In Mach's *Erkenntnis und Irrtum*, the process of idealization in the conception of theories was discussed. The resultant views, according to Cassirer, were in direct conflict with Mach's conviction that physical law has "absolutely no more factual validity than the individual facts combined."[71]

Thus it is a contradiction, so Cassirer thought, to value the individual facts only to reconstruct and interpret these facts through idealizations which ultimately "must always appear as violations of the actual facts."[72] Was it not also true, Cassirer asks, that the famous "thought experiment" (in *The Science of Mechanics*) represented a departure from the purely sensory foundations for the fundamental concepts of physics? Certainly the theory of relativity, for which Mach in many ways prepared the way, was an achievement in physical science that represented the synthetic and constructive aspects of thought. By no means can it be viewed, as Mach would have advocated, as a phenomenological reduction of science to "a mere 'rhapsody of perceptions.' "[73]

4

By the end of the century, the debates between realist and empiricist had been somewhat blunted and more sophisticated conceptions of the function and status of physical theory were being proposed. The first scientist to turn from a copy theory of knowledge to a purely symbolic theory, according to Cassirer, was Heinrich Hertz.[74] The images we utilize in theory do not denote a similarity between image and real things, but a logical relation. Thus such concepts as mass and force have no ultimate validity; they are fictions or *innere Scheinbilder* (inner illusions) created by the logic of science and subordinate to the requirements of logic.[75] As constructs, their ultimate validity or significance would be that their consequences agree with observable data.[76]

In dealing with Maxwell's new theory, Hertz was more able than others to free himself from the Newtonian frame of reference, and he did not, like Maxwell, feel the need to construct on the basis of the new equations varying models of the ether.[77] Maxwell's electromagnetic theory of light, as noted earlier, was no more nor less than the system of Maxwell's differential equations. It implied that one did not need to seek any objectivity outside of that already expressed in those relations.[78] Thus, as

Cassirer pointed out, we can readily see, on the basis of this new epistemological approach to theory, how the traditional intellectual blocks which might have impeded such a revolutionary synthesis in mechanical theory did not hinder Hertz. The manner in which the symbolic construct was integrated into his method is expressed thus: "We set up subjective pictures or symbols of the external objects, and of such a type that their intellectually necessary consequences are invariably symbols again of the necessary consequences in nature of the objects pictured . . . . Once we have succeeded in deriving symbols of the desired kind from the totality of past experience we can develop from them in a short time, as from models, consequences that would appear in the external world only after a long time, or as a result of our own manipulations. . . . The symbols of which we speak are our ideas of things."[79] No further resemblance to "things" is required except that they fulfill this practical demand.

In addition, Hertz disagreed with Mach's conception of the status of theories. Theories are not a "catalogue of isolated facts;" they exist on a different level than our sensations. They represent a free creative transformation of the data into the realm of possibility, "the expression of a highly complex intellectual process—a process in which theorizing holds full sway in order to attain to its goal through experience and therein to find confirmation or justification."[80] Cassirer states, "Thus for Hertz the fundamental concepts of theoretical physics were patterns of possible experiences, whereas for Mach they were copies of actual experiences."[81] He did not concern himself as to the metaphysical status of these theories. He did note that they are ontologically emancipated, that they are to be used as free instrumental symbolic devices for the utilization of the physicist in constructing theories in accordance with his imaginative devices, providing they allow for the empirical canons of prediction and experiential confirmation.

Like Hertz, Henri Poincaré followed a line of thought which differed from Mach's. To Poincaré, physical theory was more than a mere transcription of experience, no matter how conceived.[82] His conventionalism and instrumentalism, stressed throughout his three main works (*Science and Method, The Value of Science, Science and Hypothesis*) emphasized the truly constructive nature of theory in both mathematics and physical theory.[83] "The general laws of science—the law of inertia, the principle of conservation of energy, etc.—are neither statements about facts which can be checked by experiments nor a priori statements which necessarily emanate from the organization of human mind. They are rather arbitrary conventions about how to use some words or expressions."[84] However, theories, he went on to state, would be impotent if they were that alone. The universality of the laws of science and their necessity do not stem from the use of words. The laws, theories, and principles of science are used as a criterion of experience, not derived directly from experience. "They are not so much assertions about empirical facts as maxims by which we interpret these facts in order to bring them together into a complete and coherent whole."[85]

Pierre Duhem's thought represented to Cassirer the final step in the evolution toward a symbolic or critical conception of scientific theory. As a great contributor to thermodynamic theory and as a firm believer in the abstract mathematical formulations of the energeticists, he was inclined away from the imagists in physical theory, especially those mechanists who, like Kelvin, attempted to represent all their theories through a series of models (vortices, gears, and the like). Duhem's aversion to these pictorial methods stimulated him to carry on a rather ill-considered polemic against the British mind in science, which he characterized as being both "broad and weak," as contrasted with the French mathematical and deductive "narrow but deep" mind.

In addition to this meta-psychological nationalism, Duhem was a staunch Catholic. It was this latter allegiance which led in part to his commitment to the extirpation of all realistic interpretations of modern physical theory. Physical theory was in no sense to be identified with the real or even pictorial hypostatization of reality. His attack on metaphysical interpretations, or as he called them, "explanations," in physical theory had much in common with Mach's war on "metaphysical" or "realistic" entities. It led Frank, as well as DeBroglie to associate Mach's "economy of thought" with Duhem's view on the nature of physical theory.[86]

Cassirer, however, emphasized even more than Duhem himself that there was a real difference between Mach's economy, which was in essence a psychological crutch, and Duhem's ideal for theory as a deductive abstraction of individual laws into high-level theories characterized by their mathematical elegance and simplicity.[87] Duhem's attack on realistic explanation led him to a logical and symbolic interpretation of the nature of theory.[88] Mach's economy of thought reduced to a metaphysical sensationalism.

Perhaps the greatest exemplification of the independence of any system of symbolically expressed theories from raw sensory experience, Duhem claimed, is the impossibility of Bacon's *experimentum crucis*, by which one nonconfirming empirical discovery overthrows a physical theory.[89] Thus the experiment of Foucault with the rotating mirror, which demonstrated that light is propagated in water at a lower speed than in a vacuum, because it contradicted Newton's corpuscular theory of light and helped to initiate Fresnel's wave theory of light, was not the crucial experiment it was thought to be at that time. DeBroglie has noted that the modern quantum view of light, which does not exclude corpuscularity, has certainly vindicated Duhem's views.[90]

This advance by Duhem toward a position which not only accentuates the instrumental character of theory in the manner of Hertz and Poincaré, but which views theory as a logically independent system which initiates and conditions much of its own development, is acknowledged by Cassirer to be of greatest significance in the neo-Kantian goal of stripping physical theory of all representational imagist constructs. These imagistic principles impede the encompassing of all phenomena in a mathematical and functional system of relationships.[91] Duhem referred to this as the gradual attainment of a "natural classification" of laws

whose basic structure increasingly allows of new empirical veri-
fications and additional discoveries and accretions.[92]

One might mention, in noting the widespread intellectual
response that the neo-Kantian movement in science motivated,
the work of the French neo-Kantian Emile Meyerson.[93] Meyerson's major work, *Identity and Reality*, is contemporary with
Duhem's *The Aim and Structure of Physical Theory*.[94] Meyerson's conclusions as to the synthetic role of physical theory with
regard to the unification of common-sense experience were similar
to Duhem's. But rather than leading us toward an Aristotelian
cosmology of qualitatively ordered substances, as does Duhem,
Meyerson's idealism leads us in another direction.

Meyerson argued that the various theoretical constructs of
physics, such as space, time, force, energy, inertia, etc., which
functioned to reduce our world of experiential diversity and flux
to order, had resulted in a qualitatively empty, temporally inert,
spatially undifferentiated universe, in which matter reduces to
space and energy is conceived as a hope that "there is something
which remains constant."[95] Truly, the laws of science had overcome the raw diversity of the world. In doing so science had followed a direction given it through the human mind, which in the
end seemed to demand a universal identity, without spatial,
qualitative, or temporal differentiation.

Thus by the usual standards of common-sense thought, perception, and volition, mechanical theory is a chimera. Yet the same
process of thought applied through the medium of physical
theory, and thus having real heuristic applications, results in a
picture of the universe that is essentially inert and devoid of
experiential differentiation. From the standpoint of each of these
diverse manifestations of human thought the other results in an
incomprehensible intellectual state of affairs. This paradoxical
conclusion, the fruit of Meyerson's close contextual analysis of
nineteenth-century science and philosophy, caused him to throw
up his hands in wonder at the human significance of the scientific
enterprise.

5

Among scientists this gradual refinement of the status of scientific knowledge continued. At the same time, under the inspiration of the Kantian ideal, what one might hesitantly call the classical period of neo-Kantianism was taking root in Marburg with Hermann Cohen. Throwing off the psychologizing which characterized the thinking of Helmholtz, Zeller, and others of the neo-Kantian movement, Cohen sought to bring the movement back to the transcendental character and method of Kant's philosophy.[96] His main objection to the view in which experience is derived from our psychological organization was that Kant himself had rejected it as being an undermining factor in knowledge, since our psychological organization is an empirical and contingent fact, knowledge needing universality and necessity.[97] Cohen specifically rejected the implied receptivity and passivity found in the Kantian postulations of "the sensibility," supposedly distinct from the spontaneity implied by "the understanding." With these distinctions in mind, he felt it quite natural to expect such interest in the genetic character of knowledge which the psychological interpretation of Kant contained. Cohen centered his attention on the spontaneous activities of thought, which he conceived of in a purely idealist construction. A concept of reality outside the sphere of thought, exempt from its principles and conditions, is meaningless.[98] Reality is produced by means of thought.[99] But the important thing was to note the various directions in which thought moves in the "production of the object." This constitutes the philosophical problem, one which logic alone can trace.[100]

Here we come upon a fundamental change in Kantian philosophy. As is pointed out many times in Cassirer's and Paul Natorp's writings, and in the various commentaries on Cassirer and the Marburg Kantians, the step back to Kant was made with the thought of better understanding the kernel of Kant's critical theme—better and more consistently than Kant understood it himself.[101] In the Marburg writings the entire *a priori* structure, used to root Newtonianism in a permanent theory of reason, was sheared away. This purgation applied equally to the sensuous forms of space and time in the *Transcendental Aesthetic* as well as to the pure categories of the understanding of the *Analytic*. These constructs had provided the opening wedge for the nineteenth-century metaphysical concentration on "things in themselves." The Marburgers felt that the apriorism had prevented the meaning of Kant's epistemological method from being understood.

The "production of the object," around which Cohen centered his idealist theory of knowledge, always signified to the Marburg Kantians the "objects" or concepts of mathematics and of mathematical physics, i.e., the scientific concept.[102] Thus in contradistinction to the other schools of neo-Kantianism which flourished under such figures as Windelband, Rickert, and Dilthey, who developed theories of value and history and dealt with other such *geisteswissenschaftlichen* issues, the Marburgers followed Kant's cognitive aims into the contemporary scientific problems of the age.[103] In the writings of Natorp and Cassirer, the concrete analysis of the theories of physical science was undertaken. The forms of scientific knowledge were probed without regard to any question of origin or any concern with the "reality" of the various constructs. A vigorous attack was waged on both realist interpretations of science and the various phenomenalistic or sensationalist forms of empiricism. In order to achieve their goals of pressing forward that critical character of Kantianism, a new conception of scientific experience became necessary. Knowledge must be established to yield the necessity in experience which the theories of science show but without the *a priori* logical or psychological forms which are derived from any temporary scientific position. Hume's old challenge must be met. The necessity which any theory postulates does not derive from experience alone, yet it is not embedded within the *a priori* forms of human reason.

The solution of the Marburg school was to thrust this character of necessity or apriorism into the structure of scientific experience itself. The *a priori* does not exist before experience in a temporal sense; it can be conceived only in a logical form. Thus each theory, as it builds itself from the manifold of common sense or scientific experience, encompasses a series of judgmental orientations for the elements of the theory.[104] The *a priori* describes those logical invariants which constitute the essential structure of the theory and give it its necessary forms. As Cassirer stated: "Only those ultimate logical invariants can be called *a priori*, which lie at the basis of any determination of a connection according to natural law. A cognition is called *a priori* not in any sense as if it were *prior* to experience, but because and in so far as it is contained as a necessary premise in every valid judgment concerning facts."[105] We are not to consider the invariants of science as being permanently fixed at any one given moment. Theories change and new organizations obtain for previously known facts. As Felix Kaufmann suggested, such conceptual factors as space and time were reinterpreted in Einstein's general theory of gravitation so that the *a priori* character of Euclidean geometry, as it is found in Newtonian science, was rendered obsolete for relativity theory and a new integration of space and time was constructed.[106]

Thus, categories of space, time, and magnitude can be considered to be some of the *a priori* invariants for scientific experience and knowledge. They are meaningless terms, however, outside the specific context of the particular scientific theory.[107] This manner of analysis allows us to render the categories of thought flexible in the adaptation from more restrictive functional relationships to theories of greater and greater generality. The apparent utility of using a concept such as the *a priori* invariants of experience, Cassirer implied, is that it allows us a greater critical awareness of the character of the theory, thus opening up the possibility for a unity within knowledge based upon a synthesis of the different perspectives within the various realms of knowledge, both physical and cultural.

4

Relativity and Quantum Theory

CASSIRER, SENT TO MARBURG BY GEORG SIMMEL TO study under Hermann Cohen, the "greatest Kant scholar of the era," became absorbed with the special concerns of this neo-Kantian school. His interest in both science and history directed him to the philosophic and scientific concepts out of which grew Kant's involvement with Newtonianism. Cassirer's first systematic writings were thus concerned with Descartes and Leibniz.[1]

As he traced these historic issues in science and knowledge into the contemporary intellectual contexts, he became aware of his personal philosophical responsibilities to the philosophic tradition he had accepted. One obligation to the Marburg tradition was to demonstrate the power of Kantianism, but purify it of its time-bound conceptual associations. These included a variety of both psychological and metaphysical interpretations compatible with some aspect of the Kantian corpus but ultimately obscuring the historical significance of the great Königsberger in the minds of the Western intellectual community.[2]

Another of Cassirer's concerns was directed toward the scientific enterprise itself. Philosophy, he felt, owed a responsibility to scientists and science to aid in preventing the kinds of metaphysical hypostatizations which had plagued science since the seventeenth century and which had been repeated in differing guises with every new and radical advance. Of course Cassirer was aware that experimental scientists were not alone in their metaphysical propensities towards "explanations." Yet, even considering the high risks which any philosopher ran in falling into the same error as the scientists, the involvement was both an obligation and a natural function of the philosopher.[3]

Cassirer's first synthetic work in the philosophy of science was *Substance and Function* (1910). This volume was organized historically in its treatment of a variety of scientific concepts in their evolution to the threshold of the twentieth century. Cassirer was intent on showing that the evolution of modern science could be characterized by its attempt to avoid realistic and substantialistic conceptualizations, especially as demonstrated in the crumbling of the mechanical world view that was taking place in Cassirer's own time.[4] The inevitability with which scientists themselves came to an understanding of the functional and symbolic role of scientific theories was to him both vindication and buttress of the critical epistemology of the neo-Kantians. Just as *Substance and Function* represents the beginning of Cassirer's original contributions to modern philosophy as distinct from his more analytic historical studies, represented by *Das Erkenntnisproblem*, it also represents the first recognizable departure from the orthodox views of Cohen. Cassirer himself did not consider this to be a true break, since the great theme of the neo-Kantians expressed by both Cohen and Paul Natorp was to keep the creative insights of Kant flexible enough to adapt to new and unforeseen intellectual concerns. It was also true that Cohen's philosophy in the period subsequent to Cassirer's arrival in Marburg had undergone its own evolution. Yet Cassirer's analyses of the gradual distillation of a theory of science which eschewed metaphysical absolutes, whether mechanical or sensationalistic which had been achieved by the scientists themselves, and his rooting of this development in the history of intellectual thought brought about an exception by Cohen. As Dimitry Gawronsky reports it, the concern was with Cassirer's view that it was "impossible to ascribe an absolute value to a mathematical element, since this value is determined by different relations to which it may belong."[5]

Cohen had previously attempted to establish the infinitesimal numbers as such an absolute element. "Yet, after my first reading of your book [*Substance and Function*] I still cannot discard as wrong what I told you in Marburg: you put the center of gravity upon the concept of relation and you believe that you have accomplished with the help of this concept the idealization of all materiality. The expression even escaped you, that the concept of relation is a category; yet it is a category only insofar as it is function, and function unavoidably demands the infinitesimal element in which alone the root of the ideal reality can be found."[6]

Cassirer was insistent that the meaning of scientific concepts could only be derived by an examination of the laws which ordered and interrelated their several parts. In knowledge there could never be an absolute beginning or end. Every answer must be qualified by its relation to the existing order of categories, definitions, functional relations, etc.

By the time his next work in the philosophy of science had appeared (*Einstein's Theory of Relativity*, 1921), Cassirer's already rich humanistic interests had broadened into the area of cultural thought. *Freiheit und Form* (1916) was a wide-ranging survey of German cultural history from Luther through Schiller and Goethe to Fichte and Hegel. In 1917, in a manner reminiscent of a similar creative moment in Poincaré's life, on entering a streetcar Cassirer became aware of the conception ultimately realized in the *Philosophy of Symbolic Forms*.[7] Henceforth his views on science were developed in the context of the larger issues of a theory of knowledge which would include such concerns as language, myth, pathological thought, art, etc.

We have noted that the nineteenth-century philosophers of science struggled with the same epistemological problems as those which concerned thinkers of the late seventeenth and the eighteenth centuries, the "classical" era of modern philosophy. The final distillation arrived at by men such as Duhem, Hertz and Poincaré was consistent with the so-called critical teachings of Kant.

In these teachings, as set forth by the Marburg Kantians, a

logical distinction was to be made between our *knowledge of experience* and the supposed *reality* to which this knowledge refers. As the expectation waned that Newtonian mechanics would be the means by which all disparate realms of experience could be integrated, so too the illusion dissolved that the absolutes of this theory were rooted in a higher realm of Being. Increasingly scientists became impressed with the formal character of scientific laws as distinct from their supposed ontological status. Scientific theories were to be thought of simply as intellectual conceptualizations through which a diverse set of empirical data would be ordered.

The historic problems created by the high-flown metaphysical expectations arising from Newtonianism did not necessarily derive from the support given Newton in Kant's *Critique of Pure Reason*. If indeed it could be asserted that some of the metaphysical confidence in mechanics could be traced to Kant's authority as regards Newtonian space and time, it could be just as cogently argued, so the Marburgers claimed, that a careful reading of Kant would not impugn the philosophical integrity of his critical techniques in a world where Newtonianism was disontologized.

The important lessons to be culled from the *Critique* were not concerned with the substance of Newtonianism, nor the form of the Newtonianism, nor even Aristotelian categories. The important lesson lay in an awareness of the active, synthetic, but logical role of mind in the building up of knowledge.

1

RELATIVITY THEORY

Before the twentieth century, Newtonian mechanics was still a bulwark of the physical sciences. The problems raised by electromagnetic theories, e.g., the need to establish the ether alongside the solid, massy particles coursing their way through absolute space-time, necessitated reconstruction or at least reinterpretation. Thus, before Einstein, the Marburg Kantians could say that the realistic character of the Newtonian elements in Kant had been overemphasized. But we did not yet know the ultimate direction of physical theory. Therefore, there was nothing in Kant which directly violated the content of science, or its formal or epistemological character.

Einstein's *Theory of Relativity*, on the other hand, did raise philosophical doubts about the synthetic apriorism of Kant, now that the major Newtonian absolutes had been dethroned.[8] There were several tasks that seemed, to Cassirer, to necessitate immediate attention. First, an analysis had to be made as to the nature of the direct challenge to the historical Kant. Cassirer had to ascertain how much of the first *Critique* could be rescued for modern Kantianism. Secondly, it was important to establish whether the advances in understanding the direction and structure of scientific theory which had been achieved in the late nineteenth century in a manner consonant with the neo-Kantian epistemology could be squared with relativity theory.

On the problem of the immediate challenge to Kant there were specific issues that needed attention. The first was the adoption by Kant of Newton's use of Euclidean geometry in his spatial and temporal referential system. Whereas the creation in the 1820's by both Bolyai and Lobachevsky of non-Euclidean geometrics was merely a theoretical achievement, Einstein's use of Riemann's geometry in the general theory was a concrete envisagement of this theoretical possibility. A reevaluation and revision seemed necessary.

Second was the change in the status of matter enunciated in the special theory by which its independent and absolute nature had been modified through its equivalence with energy. In this new coordinate system the mass of a body was also a measure of its energy. The principles of the conservation of mass and of energy had now been brought into a single functional relationship.[9] This problem needed analysis with respect to the traditional Kantian position on the status of matter or substance. The final problem was the significance of the shift from Newton's and Kant's objective three-dimensional space and a distinct and objectively determined temporal division to a unified four-dimensional space-time continuum.

Concerning the status of Euclidean geometry, it was apparent that its new role merely as a special case in the total system of possible measurements established that a commitment to Euclidean geometry as an *a priori* source of our positioning of objects in space was mistaken.[10] But as Felix Kaufmann has pointed out, Kant never postulated that the possibility of a geometry different from Euclidean geometry would be a self-contradictory concept. "What had been demonstrated by the non-Euclidean geometries—provided it could be shown that they were free from contradictions—was only that the Euclidean postulate is not an analytical consequence of the other postulates. . . [Kant] had, in distinguishing the synthetic *a priori* from the analytical *a priori*, precluded such a view."[11] As it was, Kant's view of Euclidean geometry as being *a priori* for physics at all stages in this science's development—such was his commitment to Newton—was the one aspect of his theory which Einstein's use of Riemannian geometry showed to be mistaken.[12]

The new status of the matter-energy equivalency and the consequent dethronement of the Newtonian object moving independently in space and time provided some difficulties to the Kantian position. In Kant, Cassirer noted, permanence, i.e., spatial and thing constancy, is a necessary condition for the determination of phenomena as objects.[13] Substance is supreme among the concepts of pure relations. It is through this concept that a totality of sensory phenomena can be unified. Kant makes the logical demand that phenomenal nature consist entirely of permanent and lasting relations which are fixed out of the flow of change through certain universal invariants. But his demand did not necessitate the now erroneous postulation of a material substratum as the ground or foundation for the changes. ". . . the representation of something permanent in existence, as Kant occasionally said, is not identical with the permanent representation."[14] Cassirer went on to note that the transposition of the formal principle of substance into the concept of matter and the assumption of something spatially invariable were due to this historic relation of the critical philosophy to Newtonian science.

Kant's error with regard to material substance is quite similar to his error regarding the geometry of space mentioned above. In differentiating between space and that which fills it by stating that material substance can be moved independently from everything else which exists in space, he split reality into two classes of the real. The unsteady separation of the logical from the ontological was due again to Kant's overenthusiastic commitment to Newton as representing science at all future stages of growth. Thus the distinction between the form of the sensibility and the matter that fills it as established in the "Transcendental Aesthetic" and the "Transcendental Analytic" was gravely undermined in the general theory of relativity.[15]

Kant's position with regard to space and time suffers less, as Cassirer pointed out, than the aforegoing aspects. There was in the critical teachings a purer strain which was consistent, as even Einstein recognized, with the epistemological implications of relativity.[16] As early as 1758 Kant had set up a principle of the relativity of all motion, only to fall back in 1763 and 1769, on Euler's authority, on the concepts of absolute space and time. It was Cassirer's claim that in Kant's transition from physics to "transcendental philosophy" in his Inaugural Dissertation of 1770, realistic questions were again reshaped in terms of epistemological issues. Here space and time are studied for their significance as concepts in the structure of empirical knowledge. They are not "independent objects, which are somehow present, and which we can master by experiment and observation, but 'conditions of the possibility of experience.' "[17]

Space and time make possible the establishment of objects; they are not objects distinct from other things. In establishing the existence of things space and time achieve a certain "objectivity" in that they lead us to making judgments about things. The meaning of space and time in the structure of science is given by the judgments that are built out of their relations between the other conceptual entities of science. What they are as things in themselves Kant established as an unintelligible question. "Space and time possessing an *existence* separate from empirical bodies and from empirical events, are rejected as nonentities, as mere conceptual fictions (*inane rationis commentum*). The two, space and time, signify only a fixed law of the mind, a schema of connection by which what is sensuously perceived is set in certain relations of coexistence and sequence."[18]

Thus when Einstein stated that the theory of relativity took from space and time the last vestige of physical objectivity, it coincided with critical idealism in its purest form. As Max Laue pointed out,

> The boldness and the high philosophical significance of Einstein's doctrine consists in that it clears away the traditional prejudice of one time valid for all systems. Great as the change is, which it forces upon our whole thought, there is found in it not the slightest epistemological difficulty. For in Kant's manner of expression time is, like space, a pure form of our intuition; a schema in which we must arrange events, so that in opposition to subjective and highly contingent perceptions they may gain objective meaning. This arranging can only take place on the basis of empirical knowledge of natural laws.[19]

One might suggest that where Kant approached Leibniz on the ideality of space and time he adhered most consistently to the critical theme.[20] Because of this vision of the relational and functional view of space-time, we are freed from absolutes of objectivism inherent in Newtonian realism as well as from the superficial ascription of subjectivism to the relativity theory, since no one coordinate system has priority over any other moving system. It is also relevant to emphasize that Kant's space and time of pure intuition are not psychologically or perceptually sensed—as Moritz Schlick had on one occasion claimed. Cassirer noted that the supposed subjectivity of space-time in Kant is not a psychological subjectivity, but is a logical or transcendental subjectivity that provides for the objective and mathematical construction of Newton's space-time. Einstein took this earlier "objectivity," which is inherently limited to one coordinate system (Newton's), and universalized it so that "objectivity" now refers to relations between diverse sets of space-time systems.[21]

The space-time continuum has now a wider conceptual meaning, since all things moving in diverse systems at diverse speeds are correlated to each other through the laws of relativity. These mathematical equations conceptually order a number of heretofore disparate qualities of experience, including the previously enigmatic equivalences of inertial and gravitational motion now made intellectually significant in addition to being empirically established and theoretically noted (as it had been in Newton).

2

We next turn to Cassirer's examination of the import of relativity theory for the then current functional and symbolic interpretations of scientific knowledge. The historic utility of Kantianism in establishing canons for examining the metaphysical or sensationalistic reductions of knowledge in the eighteenth and nineteenth centuries is recognized as a by-product of Kant's systematic endeavors. His critical insights were applied in the context of the disintegration of two great intellectual traditions—the era of the empiricist-rationalist debates of the eighteenth century and the period of the decline of Newtonianism in the nineteenth century. Einstein's theory was a revolutionary if classical scientific accomplishment (as compared with quantum theory). Its success and failure depended not on philosophical and epistemological considerations but on logical, formal, and empirical determinations. The ultimate question was not whether relativity theory would render the instrumentalism and conventionalism of the Marburg school incorrect or vice versa, but whether a theory such as Einstein's, having such a potentially enormous impact on our understanding of physical experience, could lend indirect support to the meta-theoretical contentions of the critico-idealist position on the nature and meaning of scientific knowledge.

This latter problem was not uppermost in Cassirer's thought when he first undertook his analysis. Before 1920 relativity was still a somewhat controversial and to some a tenuous conceptual program. Cassirer's historical analysis of science had, in addition, secured for him an assurance of the rightness of his criteria for both the form and import of scientific theory. His analysis then was directed toward securing the place of relativity theory as well as adding one more significant confirming instance to his own epistemological position.

The first test which relativity theory had to meet with regard to the critico-idealist position was what might be called the test of lawfulness and inclusivity. The import of this logical demand was that relativity theory conform to the traditional goals of science as regards objectivity or public knowledge. It had to refute the charge of personalistic subjectivity directed against it by the many naive commentators of that era. Also, it had to persuade those who viewed physical science only in the traditional Newtonian mechanical garb it had worn for so long. Finally, relativity theory needed to exemplify the logical demand for unification, by which we are able to bring a variety of experiences into logical correlation with each other as well as enabling us to return to perceptual experience and demand from nature further control and mastery.

The second criterion was that relativity theory ought to exemplify and demonstrate that quality of thought with which science substitutes logical and mathematical invariants for substantial absolutes, as for example Duhem had demanded of physics in his critique of the British model builders.[22]

Cassirer felt that relativity theory strictly conformed with these epistemological canons. For example, as regards the first criterion, in the older mechanical tradition, objectivity was believed to be achieved if one could establish empirical determinations of a body as it moved through space and time with a certain velocity. It was believed that both definitive and universal information could be had, through the processes of measurement and analysis, that would encompass all bodies within our universally objective system of reference. The theory of relativity, by making the process of measurement dependent on the motion of the entire coordinate system and by making measurement in any one system merely a relative determination, not an absolute one, had seemed to some to dissolve the very structure of scientific objectivity. On the contrary, Cassirer claimed, Einstein had allowed us to see our coordinate system in the context of a larger consideration, namely the state of motion of this reference system from which, in turn, particular space-time values are being taken.[23] The relativization is thus only logical and mathematical and is not dependent on the idiosyncratic sensory observations of the individual.[24] Objectivity is placed in the context of a new invariance, which makes the older Newtonian space-time continuum a special case insupportable in terms of the new theory, but convenient for use in most cases.

In establishing the speed of light in a vacuum as a new and basic invariance, Einstein refers us back to the role that the velocity of light plays in all of our physical-time measurements.[25] Thus the length of a body, its volume, form, energy, and temperature are to be considered according to the choice of the system of reference in which the measurements have taken place. This is derived from the formulae of the Lorentz transformation. The constancy and unity of nature as a conceptual foundation are saved, despite variations in the measurements of space and time in the different coordinate systems, by the uniform magnitudes attached to the velocity of light, the entropy of a body, its electrical charge or the mechanical equivalent of heat, regardless of reference system.[26]

The advance which has taken place lies not in the realm of the clarification of experience nor even in the strict adherence to scientific regard for facts, evidence, and experiment:

> The postulate of the constancy of the velocity of light and the postulate of relativity show themselves thus as the two fixed points of the theory, as the fixed intellectual poles around which phenomena revolve; and in this it is seen that the previous logical con-

stants of the theory of nature, *i.e.*, the whole system of conceptual and numerical values, hitherto taken as absolutely determinate and fixed, must be set in flux in order to satisfy the new and more strict demand for unity made by physical thought. . . . in the words of Goethe, experience is always only half experience; for it is not the mere observational material as such, but the ideal form and the intellectual interpretation, which it is given, that is the basis of the real value of the theory of relativity and of its advantage over other types of explanation.[27]

3

The general theory of relativity exemplified another traditional quality of the scientific quest for theories of ever more unifying capacities. Always, stated Cassirer, science searches for lawlike functional relations so that the scattering and fragmentary aspects of experience can be regularized by abstract relations which can theoretically hold diverse phenomena. Thus individual tastes, smells, auditory and visual sensations are replaced by a set of constants which relate the various sensory aspects of an object to a series of mathematically expressed functions. Thus the "nature of a body, in the physical sense of the word, is determined not by the manner of its sensory manifestation but by its atomic weight, its specific heat, its exponents of refraction, its index of absorption, its electrical conductivity, its magnetic suscepti-bility."[28] Each of these individual determinations is achieved through certain constants which link and order the particular case with a vast number of other particulars through a law which remains invariant.

The course of scientific theory is thus characterized by the attempt to link each of these constants through higher level theo-ries, by the development of even more general constants in which the prior regularities become expressions of a more particular order. Thus, as Cassirer pointed out, Dulong and Petit in 1819 found a relation between the atomic weight of an element and its specific heat: the specific heat of an element is inversely pro-portional to its atomic weight. Richards later confirmed this by deriving this relationship from the kinetic theory of heat. Even later it was found that certain solids of low atomic weight show a deviation in the product of their atomic weight and specific heat, which according to the original Petit and Dulong law should have been identical for all elements. Einstein finally resolved this by transferring the problem to the context of quantum theory as applied to the thermics of solid bodies.

Further, an important unity between diverse fields was achieved when Maxwell established a relation between the optical and electrical behavior of certain substances. The fact that exact empirical confirmation was not forthcoming in all instances did not limit the explanatory power of the theory. But a search ensued, and confirmation was finally achieved by a more careful refinement of the dielectrical constant through the electron theory of dispersion.[29]

A most important example of this trend is shown in the general theory of gravitation by the connection which Einstein theoretically established between the weight and inertia of masses. The equality was established by Newton and later confirmed decisively by Eötvös and Zeemann, with the torsion balance.[30] At the very beginning Einstein was concerned with Newton's method of registering this equivalence without giving it theoretical interpretation. The result of Einstein's ultimate incorporation of this empirical fact into a theoretical context was of course revolutionary. Although Newton attempted to avoid metaphysical ascriptions to the physical processes he described in his equations, Cassirer conlcuded that the very structure of Newton's conceptualization doomed his attempt as regards gravitation. Newton's mere "description of phenomena" without theoretically uniting the concepts of "gravitation" and "body" necessitated an interpretation of gravitation as something external and alien to matter. Weight was thus a universal but not an essential property of matter.[31] In effect Newton had produced in this postulation of gravity a truly metaphysical concept.

With Einstein a numerical proportion found universally between inert and heavy masses became the expression of physical equivalence or likeness. It is the same quality of body which is expressed according to circumstances as inertia or weight.

In electromagnetic theory the identity of light waves and electrical waves is also achieved in the representation of these diverse experiential phenomena through the same equations. The logic of scientific thinking, Cassirer argued, demands that we increasingly explore the possibility for expressing such physical equivalencies in terms of formal or mathematical relationships of identity. Einstein's theoretical achievement thus stands at the end of a long line of development which began with Sadi Carnot's break from a theory of substances to a physics of principles, as exemplified in his thermodynamics. The equivalence and unification which Einstein found are an equivalence in the area of judgment. Thus if an acceleration occurs, an observer can interpret it as the effect of a gravitational field or as an accelerated motion within a different system of coordinates. Thus, "to attain a universal theory of gravitation we need only assume such a shift of the system of reference and establish its consequences by calculation."[32] The general theory of relativity represents the most encompassing postulate of dynamics in its establishment of the invariance of certain magnitudes and laws with regard to all transformations of systems of reference. It is perhaps the culminating theory in a long historical process, so Cassirer concluded, during which the unifying tendencies within the physical sciences have always been evident.[33]

It is important to ask the philosophical significance of this shift from one set of invariants or absolutes to another. What is the status of the invariants? Here the usefulness of Cassirer's deontologized reinterpretation of Kant's synthetic *a priori* is realized. The invariants of every discrete theoretical system function in a purely logical manner by providing the necessary premises by which numerous particular judgmental orientations regarding concrete experimental and experiential issues will be decided[34]

Thus the *a priori* does not denote temporal or metaphysical priority, before experience. Rather it signifies the manner in which subsequent deductive conclusions are always predicated on certain logically prior assumptions, axioms, or in this case invariants of the theory.[35] Judgments and decisions that are made with respect to similar observational data are invariably different in kind as the more general considerations of the theoretical structure shift from one set of invariants to another.

The evolution of physical theory therefore does not represent a steady additive quality. The shift from one *a priori* logical structure to another gives the entire superstructure new significance and meaning. Thus, though our present set of conceptual measuring devices deals in the most absolute manner with the phenomena of experience insofar as the flux of things is given stability and permanence, we know that this intellectual state of affairs is not fixed. Both from past experience and from the very functional nature of our present theoretical absolutes, we know and expect them to be supplanted eventually by an entirely new set of concepts.

It is not true, Cassirer stated, that the evolution of science demonstrates a mere shifting back and forth from one relatively true theoretical center to another, though this phenomenon has a certain historical justification, e.g., Huyghens and Newton—thermodynamic theory—Heisenberg and Schrödinger.[36] Nor can we state that the uncovering of new facts or observations which do not agree with prior theoretical articulation renders a theory a fatal blow.[37] On the contrary, Cassirer noted, "thought has found in them [facts] a new point of leverage, around which moves henceforth the totality of empirically provable 'facts.' "[38]

What has occurred in the overall development of scientific theory is that each new set of theoretical invariants represents a progression toward not only the ever increased infusion of thought into scientific procedures but also toward the gradual encompassing of the particulates of experience. It is not that the most generalizing conceptions attempt to eliminate the specificity and concreteness of the phenomena of experience. The more we are able to encompass into one large framework with its highly abstracted invariants—and it is the role of epistemology, Cassirer argued, to delineate the logical and therefore essentially ideational character of these invariants—the more the entire structure of physical experience will be unified through a highly articulated set of logico-mathematical relations. In the context of the exact sciences this constitutes the paradigmatic symbolic form in which thought is expressed.

4

This brings us to the other dimension of the development of physical theory. For if it is true that there is a direction toward more encompassing theories, each with its particular invariant assertions, and if the evolution of theory displays, as Cassirer claimed, a steady accretion of ideational form, it is necessary to explore the particular character of this development. This process does not take place in a unilinear manner; it is opposed by a divergent and recalcitrant tendency in thought. There is, Cassirer stated, a dialectic around and through which progress takes place because theory, through its logical and rational power, must overcome strong imagistic and substantialistic qualities of thought. The road to scientific objectivity has been a struggle, upward and beyond the common-sense categories of perceptual experience.

> When asked wherein the strictly objective factor in nature consists, the modern physicist can only point to the universal constants arrived at by investigation and, on the other hand, trace the road leading from these universal constants down to the individual constants, the particular, thing-constants. At the summit of his system stand certain invariable magnitudes, such as the velocity of light in a vacuum, the elementary quantum of energy, etc., which are free from any merely subjective contingency, since they prove to be independent of the standpoint of the individual observer. The process of physical objectivization is an ascent from mere material constants, from the particularity of thing-unities to universal unities of law.[39]

The example given by Planck in his survey, "Of the Origin and Development of the Quantum Theory," is representative of this intellectual process.[40] Planck's first ideas related to Kirchhoff's law of heat radiation, in which under certain conditions heat radiation was unaffected by external properties of bodies so that a universal function depending only on temperature and wave length was achieved. Other related constants were soon noted by Stefan and Boltzmann, and Wien formulated the law of displacement, which established a constant now defined as the product of wave length and absolute temperature. These diverse ideas, though still in the process of development, were finally joined together in 1900 by Planck in his quantum theory. This theory linked the empirically determined constants of Stefan and Wien with two fundamental magnitudes: the elementary energy quantum and the mass of the hydrogen atom. These "thing" concepts from several fields of investigation were now linked together through a new and encompassing theoretical idea. To Cassirer, the "thing-concepts can at most connect, but the physical concepts combine—the thing concepts create a coexistence of properties as mere particulars, but the concepts of physics go on to posit truly universal unities."[41]

One can exemplify the manner in which Einstein's theories furthered the process of idealization—as identified by the neo-Kantians and achieved without the metaphysical errors of previous interpretations of scientific theory—by noting how those three fundamental qualities of physical experience, space, time, and substance, are now conceived. Their very character has been shorn of the last vestige of intuitive forms of representation. The theory of substance which in the course of nineteenth-century physics had been bifurcated into a dualism of mass and energy was united in the special theory. Both conservation theories logically give way when "the inertial mass of a body is not a constant, but varies according to the change in the energy of the body. The inertial mass of a system of bodies can even be regarded as a measure of its energy."[42] The only invariant is the factor of time as it relates to the Lorentz transformation.

The idea of substance in its traditional intuitive, atomistic, or even Newtonian envisagement disappears. It cannot be represented pictorially but only through the relationships which are given meaning in the equations of relativity. "Here the substantial is completely transposed into the functional: true and definitive permanence is no longer imputed to an existence propagated in space and time but rather to those magnitudes and relations between magnitudes which provide the universal constants for all description of physical process. It is the invariance of such relations and not the existence of any particular entities which forms the ultimate stratum of objectivity."[43]

Space and time likewise cannot be conceived of as individual entities, each with its own unique objective character. They do not exist independently of each other. The equations relative to one ultimately define the conditions of the other. "All physical, field phenomena are expressions of world metrics,"[44] stated Cassirer, expressing again how close modern physics is to the geometrical method of Descartes.[45] All spatial and temporal values are now interchangeable in this continuum; temporal values of past and future are cancelled out by arbitrary spatial directions established by definition. Sensuous intuitive concepts of space and time are absorbed into a uniform and homogeneous mathematical system. By eliminating qualitative imagery, by surpassing the most extreme demands of Galileo, physics fulfills itself by being enveloped in the Pythagorean garment of number.[46]

Einstein's theory of relativity revolutionized the discipline of classical mechanics. It also enriched the meaning of mechanics by providing a supplementation to its postulates in an area where the theory was breaking down. It did this with a radically new interpretation of the traditional absolutes of space, time, mass, and velocity, giving them a new meaning of the most general and abstract nature. And yet the logical character of this new theory was completely consistent with the conception of physical theory that had been developed through the criticism attendant to the disintegration of the ontological claims of Newtonianism. The symbolic view of physical theory could thus be retained amidst one of the greatest upheavals in the history of science.

5
QUANTUM THEORY

Quantum theory did not burst upon the intellectual world with the same power as did Einstein's relativity, perhaps because of the contrasting character of the two theories. Einstein's reconstruction of Newtonian mechanics derived from several powerful if implicit criticisms of Newtonianism as exemplified in the Michelson-Morley experiments and the Lorentz transformation. By a general deductive reconsideration of the basic postulates of Newtonian mechanics in the light of the new experimental and theoretical evidence as well as the established experimental laws of the earlier theory, new conceptions such as relativistic mass and a new conservation of momentum were introduced.[47] More than any particular experimental results which later buttressed the theory, it was the theoretical grandeur involved in these basic redefinitions of concepts in the light of the newer invariants that focused the attention of the intellectual community on Einstein.

By contrast, Planck's introduction of the quantum followed a long and problematic concern with the theoretical difficulties involved in black-body radiation, leading from Kirchhoff and Boltzmann through Wien and finally to Planck himself, who was long a strong adherent of the mechanical world view and was thus most reluctant to recognize the revolutionary philosophical significance of his own theorem. But with Einstein's association of constant "h" in the experimental laws concerning the photoelectric effect, and its consequent extension to other experimental phenomena, such as the line spectra of elements, the specific heat of solids, and later, through Bohr's, description of the structure and functions of the atom, the importance of this invariant measure of energy has grown to its present significant status.

The philosophical attention given to this theory has in spite of its indisputable heuristic utility gone far beyond that which attended the birth of electromagnetism or even relativity, both of which forced modifications of our traditional mechanical assumptions concerning the nature of matter. Perhaps primary among the various reconsiderations which quantum theory made necessary was the logical integrity of the atom, which, as Cassirer pointed out, had held firmly from its earliest postulations by the Greek and Roman atomists, through Boyle, Huyghens, and Gassendi into the nineteenth century. The concept's various confirmations, as in the statistical theory of heat and ultimately in observations, such as those of Brownian movement, had even persuaded phenomenalistic scoffers, such as Ostwald and Mach, who followed Rankine's theoretical criticisms of such "occult" entities as the atom, into accepting its apparent "reality."[48] As a result of the experimental and mathematical interpretation given by quantum theory, the atom had now turned into an indistinct, conceptual blur. It had not only lost its materiality, but also its sensory distinctness and particularity as a physical entity. Now, it was "either—or, somewhere, having a certain probability."

The attempt to place this area of investigation under certain regulative conditions of thought so that research could be directed into further avenues of development had resulted in a theoretical shift. The older (Laplaceian) objectivity—where, given a body moving through space and time with a given velocity and with specified forces acting on the body, the location and velocity at any future time might be predicted and measured—could no longer exist in quantum mechanics as a standard of experimental and theoretical enquiry. A new canon of objectivity was now reified into conceptual use as a criterion for evaluating the results of experimental enquiry.

Only probabilities could now be inserted into the description of the action of large numbers of electrons. To speak of the position and momentum of any one electron was prohibited by the basic conditions under which experimentation, observation, and prediction took place. The depths to which this theoretical alteration was rooted could be seen by contrasting this new probabilistic description with that of statistical mechanics. As Northrop points out, in Newton's and even Einstein's theories the concept of probability is "restricted to the epistemological relation of the scientist in the verification of what he knows; it did not enter into the theoretical statement of what he knows."[49] The observation and prediction of the motion of atoms in traditional mechanics were derived as a means of experimental convenience from large numbers given in a probability relationship. In theory, the dynamics of the state of the system could be predicted, therefore, so too the state of the individual components, given the existence of techniques for observation. In quantum theory the probability relationships enter "theoretically and in principle; they do not refer merely to the operational and epistemological uncertainties and errors, arising from the finiteness of, and inaccuracies in, human behavior, that are common to any scientific theory and any experimentation whatsoever."[50]

In spite of the great changes in the formalism of the theory as compared with classical explanations, the predictive, experimental results are wholly within the compass of traditional requirements of the scientific method as concern prediction and confirmation. As Philipp Frank notes, "The initial arrangements and the results can be described in the language of classical physics because we have to do with familiar mechanical and optical objects like medium-sized bodies, bright fringes, and point-like scintillations on a screen, etc. However, the principles from which we can derive the connection between initial arrangements and results are, as we have learned by now, not principles that contain concepts that are familiar from Newtonian physics; they cannot be expressed by describing trajectories of particles, propagation of waves in a medium, or similar concepts."[51]

In the mathematical formulation of DeBroglie's wave mechanics the amplitude of the wave (itself formulated in probability terms) is uniquely determined if the initial experimental arrangement is given.

> We know only how many scintillations will appear on the average per unit of time in a certain region of the screen, but we can never predict at what exact location and at what precise instant of time a particular scintillation will occur. Therefore, we must say that there

is no causal law that allows us to compute from an initial arrangement the precise location of any single scintillation on the screen. We can compute the exact value of the wave amplitude at every point, but this amplitude is not observable; it is only connected with phenomena by the operational definition of the amplitude as proportional to the probability of a particle's being in a certain region. To be exact, we should not speak of a "particle's being in a certain region," because there are no particles. We should speak of the probability that scintillations occur in some region of the screen or about the frequency of scintillations in this region. Thus, the mathematical formulation, including the operational definitions of the symbols in this formulation, provides us with rules that connect the initial arrangements with "observable results."[52]

6

Cassirer's *Determinism and Indeterminism in Modern Physics* was written in 1936. It therefore confronted a theory which at that time had received only partial articulation. His analysis of both the philosophical and scientific issues then being raised was predicated on the existent state of knowledge in the discipline. Cassirer's death in 1945 ended his plans to add a chapter that would bring the argument into focus with recent research. His analysis of the themes purveyed by the philosophical detractors of quantum theory, as well as the defense he proposed, as Margenau notes, are completely consonant philosophically with what has taken place intellectually since its publication.[53]

Perhaps the most traditional philosophical concern with the meaning of quantum theory was exemplified in the writings of W. Nernst, who stated that modern physics now revealed an unexpected parallelism between theology and physics. Traditional theology always claimed that cosmic happenings were determined by the will of God and that man was forever denied this ultimate knowledge. Man, through his intellectual capabilities, could approximate an awareness of the grosser regularities of reality, but not of their complete determination. Nature, too, as revealed by quantum physics, so this argument went, withdrew the possibility of such determinations. We would therefore be forced to accept only a partial articulation of reality and alternative versions of what the partially known real world of matter was ultimately like.[54]

Werner Heisenberg felt somewhat the same way when he first postulated his indeterminacy theorem. He stated that there existed limits to man's knowledge of the precise position or velocity of a particle or of the exact determination of the wave- or particle-like qualities of electrons or photons. Quantum theory therefore was to be proposed only as an instrumentality of thought whose function lay in regularizing these ambiguities into a principle of scientific investigation. However, because of the very peculiar character of the conceptual entities to which the law applied, as well as its unusual form, Heisenberg felt that it could not take its place in the traditional pantheon of scientific theories and indeed might have to be excluded from the realm of strictly causal and deterministic theories in the structure of science.[55]

Heisenberg felt that, scientifically, quantum theory presupposed few difficulties when one is concerned with a strictly empirical description of the facts involved in the experimental process. But as soon as one attempts to rise from the arena of sensory experience to a theoretical discussion of the events in question, especially with regard to the atomic particles, the mathematical schema fails to correspond to or complement the language used to describe the experimental events. Heisenberg's interpretation of this lack of logical commensurability is as follows: "In the experiments about atomic events we have to do with things and facts, with phenomena that are just as real as any phenomena in daily life. But the atoms or the elementary particles themselves are not as real; they form a world of potentialities or possibilities rather than one of things or facts."[56]

As F. S. C. Northrop claims, the classical mechanical model, with its explicitly defined conception of causality, is one with the demands of knowledge for a complete description of the object of modern scientific knowledge. This demand cannot be maintained in the "state" descriptions of quantum theory; therefore we are given an incomplete description of nature. Presumably, behind the phenomena of quantum theory there are noumena, whose behavior is hidden from man's enquiring mind. Echoing Einstein's comments in reply to Niels Bohr's defense of the theoretical form of the principle of complementarity, Northrop states, "Experiments on black-body radiation require one to conclude that God plays dice."[57]

Einstein, reflecting the earlier concerns of Planck, and the later hesitations of DeBroglie over the form of the theory, stated his reservations concerning the incomplete symbolic representation of reality.

> But now I ask: Is there really any physicist who believes that we shall never get any inside view of these important alterations in the single system, in their structure and their causal connections, and this regardless of the fact that these single happenings have been brought so close to us, thanks to the marvelous inventions of the Wilson chamber and the Geiger counter? To believe this is logically possible without contradiction; but it is so very contrary to my scientific instinct that I cannot forego the search for a more complete conception.[58]

It is true, Cassirer stated, that quantum theory had deviated from the traditional form of scientific assertions and perhaps traditional requirements for supposedly universal canons of objectivity. But to argue for one kind of objectivity is to demand too much of the specific form of a scientific theory. If one adheres to the specific limitations set up by the uncertainty relation, precisely definable functional relations will exist between the theory's various elements.[59] Therefore it could not be claimed that quantum theory had brought into nature a wholly new set of principles.

It is true that these relationships will not be those assigned through classical mechanical theory, but as Dirac has shown, quantum mechanics does formulate its laws in such a manner as to allow specific determinations to be made without ambiguity under any given experimental conditions.[60] Therefore, merely because the specific form of these laws is altered so that the laws interpret the system in a statistical manner, nature does not become incomprehensible. In addition we are not warranted in ascribing to nature a reality which is inherently probabilistic or infused with "potentiality."

> If the interpretation associated with the Psi-function is a statistical one, then all predictions based exclusively on that interpretation must also be statistical, and cannot be predications of nonstatistical properties to individuals. There is therefore no warrant for the conclusion that because quantum theory does not predict the detailed individual behaviors of electrons and other sub-atomic elements, the behavior of such elements is "inherently indeterminate" and the manifestation of "absolute chance."[61]

Cassirer felt that there was no philosophical need to consider the abandonment of the principle of cause and effect, but perhaps it was necessary to revise our traditional views on the conceptual status of substance and attribute. With quantum theory we cannot know in an unqualified and absolute manner that that which is under investigation is under all conditions and circumstances to be thought of as either wave or particle. An electron or photon can be described as either, under differing experimental and observational circumstances. But there is no experiment by means of which both the wave and particle nature of light can be demonstrated simultaneously.[62] But this difficulty does not herald the breakdown of lawfulness in nature.

> The abandonment of absolute determination restores the highest degree of relative determination of which physical knowledge is capable. For if, in the determination of a physical system, we admit only such elements of determination as satisfy the conditions expressed in the uncertainty relations, if we are satisfied with maximum observations — that is, with the largest number of independent compatible observations — we can bring these into a sharply defined relationship with each other. We can then establish

the principle that when a maximum observation of a physical system is made, its subsequent state is completely determined by the result of this observation and this principle can be used as the axiom to express what we are to regard as the "state of a system" in the sense of atomic physics.[63]

As a result of this view it becomes necessary to change our traditional ideas as to the meaning of the reality of a thing in the context of modern physical theory. Traditionally, a real object was established epistemologically by noting its place, its spatio-temporal existence. As late as 1910 Paul Natorp, a fellow Marburger, stated that the existence of a thing and our knowledge of it depended upon our being able to locate the object at a determinate place and time.[64] Moritz Schlick, having had to admit the impossibility of such a determination in space, thought that an unequivocal time determination was to be regarded as a necessary criterion of reality.[65]

These criteria, Cassirer stated, need to be abandoned, even at the risk of once more departing from some of the time-honored Kantian constructs. It is not the Kantian epistemological position that we are abandoning but merely constructs that were too closely joined to a particular state of our scientific knowledge.

In quantum theory we cannot establish the existence of a mass point to be simultaneously in a given place and at a certain time. In wave mechanics we think of a "particle" as being widely diffused through space. There is a kind of "presence" to its mass and charge. Rather than the Keplerian orbits of the classical electron these orbits now are an indefinite "charge cloud."

There now exist two pictures described by two different languages for the results of our experiments and observations. Neither of these state systems is ever defined clearly or exhaustively.[66] If we use one measuring device, a wave or particle conception of, let us say, light is revealed; another set of empirical conditions dims the first picture and another is given to us in its stead.[67] Thus in Bohr's principle of complementarity certain experimental arrangements allow the scientist to define the position of atomic objects, others to define the objects' momentum. Specific physical operations thus can lead to the assignment of specific values as to the spatial coordinates or velocity component of this object. If the position of the object can be derived through the experimental arrangements, then conclusions about the diffraction pattern which will be produced by the scintillations on the screen can be predicted. If, on the other hand, the momentum is ascertained through the movement of a diaphragm, it is impossible to define the position of the particle, although other predictive results (velocity, motion of diaphragm) now become possible.

While this may not be fully satisfying theoretically, as compared with the clear-cut specificity of Newtonian mechanics, there is a gain for epistemology in these special sets of circumstances. Heisenberg, who first set forth the uncertainty relation, early realized the danger inherent in the attempt to make a pictorial or intuitive representation of a set of mathematical relations.[68] In quantum theory, Cassirer pointed out, both representations are elevated in principle to a set of theoretical correspondences so that combined they furnish us with a precise conceptual formulation of the observable phenomena, such as within the structure of the concept of the atom.

This implicit critique of pictorialism in physical theory is well within the neo-Kantian tradition, wherein theories are viewed as creations of thought whose function is to order experience and to conform it to the intellectual requirements of lawfulness. The traditional views of the atom and the material point as logical constructs claimed as their historical *raison d'être* for science the potentiality to free thought from the flux of sensory experience and yet not force physics to leave the realm of common-sense imagery in retaining this symbol of permanence.[69]

But now we have a theory which cannot be accounted for in terms of ordinary experience. When in our experiments we produce a diffraction pattern which depends on the behavior of a host of electrons, it is impossible to assert the status of one lonely mass point. The diffraction pattern produced by an emitting body is the creation of "probability waves" whose theoretical state can only be described in a statistical manner.[70] It thus becomes impossible to conceive of the empirical phenomena in terms of the actions of fully determined rigid material particles.[71]

Quantum theory as compared with relativity theory has thus followed a different path of conceptualization, partly brought about by the difficulties inherent in the interpretation of the empirical phenomena encountered, and partly by the theoretical sophistication gained through the experience of the scientists of earlier generations. One might not be happy with the present theoretical state of affairs in quantum mechanics or even expect the present formulation to approximate the theory's final state.[72] That it fulfills the traditional requirements of theory, however, cannot be doubted.

As Cassirer put it:

> Modern quantum mechanics thus tended more and more to begin not by positing definite realities, which are subsequently brought into relation with each other, but rather by choosing the opposite path. It starts out with the establishment of certain symbols expressing the state and the dynamic variables of a physical system. From these, on the basis of definite axiomatic presuppositions, other equations are derived, and physical consequences drawn from them. At first it is not necessary to dwell on the exact significance of the symbols in a particular case. Only at a later stage of consideration are the representations of the abstract symbols examined — in other words things and attributes are examined [and postulated] which satisfy the rules valid for the interrelationship of the symbols.[73]

It was his opinion then, that the heuristic value of certain traditional scientific cornerstones must be questioned, if not in this special context abandoned. For long, they were pivots around which experiment and theoretical system building interacted and expanded to encompass large new areas of experience. But now that these concepts—position, velocity, mass—of an individual electron no longer have empirical meaning, they may have to be excluded from theoretical physics, Cassirer argued, important and fruitful as their function in the history of science may have been.[74]

7

The direction which quantum theory has taken is, as we have just noted, somewhat different from that of relativity. The latter might be seen as a traditional nineteenth-century theory in that it not only completed the Newtonian system but its basic conceptual formalism was consonant with the traditional Newtonian demands for complete objectivity once the latter was sheared of its traditional imputation of giving absolute status to the coordinate system of the earth and the fixed stars. Newtonianism could then be absorbed into the more general relativity theory as a *special* but not logically derived case of the latter.

Quantum theory came upon the scene as a heuristic theory aimed at correlating a number of experimental peculiarities associated with black-body radiation. As this whole line of research opened up a number of new areas of experimentation, quantum theory became a synthesizing element in bringing methodological rigor to various areas of atomic theory, optics, and heat. The fact that the theory was able to bring order and generality to such a disparate set of experimental realms exemplified its role in fulfilling the now traditional logical canons of scientific theory.

In addition, and perhaps more characteristically, the theory fulfills a long-standing neo-Kantian expectation that the traditional historical involvement with images, models, or even objects in science would be replaced by theories couched in more abstract and relational terms, and thus having the additional virtue of being free from metaphysical interpretations. This freedom from the philosophical requirement for "explanations" enhanced its power to deduce and derive new implications from the basic experimental matrix of experience.

It was this change in the character of the theory, especially its probabilistic avoidance of strictly mechanical conceptions of causality, which bedeviled so many thinkers. Those philosophers and scientists who followed Niels Bohr believed that the principles of causality and the amenability of natural processes to lawful descriptions did not have to be surrendered.[75] By not linking causality with any specific set of naturally observed sequence of events, as Jeffrey has attempted, or by not envisioning causality in strictly Newtonian mechanical terms, the integrity of these principles was preserved.[76] But it was preserved, in Cassirer's interpretation, because scientists perceived the active constructive role of thought in the pursuit of general principles.[77] It is this lack of realistic objectivity which concerns Northrop, causing him to conclude that quantum theory has brought about a form of "ontological mechanical causality" due to the ontological role which probability plays in quantum theory, as compared to its "theory of errors role" in the Newtonian and Einsteinian contexts.[78] It is but a short step for Northrop to assume that since the theory cannot be expressed in a completely deterministic manner —and that its very symbols represent statistical expressions—the symbolism reflects an ontological characteristic of the nature of reality itself. Indeed, perhaps the inscrutability of physical reality has some relevance in understanding freedom and indeterminism in the behavior of human beings when the latter are examined from the standpoint of physics: "potentiality and the weaker form of causality hold also for countless other characteristics of human beings, particularly for those cortical neural phenomena in man that are the epistemic correlates of directly introspected human ideas and purposes."[79]

But as Nagel phrases it, somewhat similarly to Cassirer, one cannot ascribe any special form to causality nor can one claim a metaphysical mystery to exist in the supposed conditional character of reality merely because one form of causality is not universally applicable. The causal principle is "no law of nature and has no identifiable descriptive content. On the contrary, the principle functions as a maxim, as a somewhat vague rule for directing the course of inquiry, as an injunction to interpret and organize our experience in a certain manner. . . . But there is no doubt that to the extent the principle is accepted as a regulative principle for inquiry, it functions as a maxim and specifies a criterion for what is to be counted as systematic knowledge."[80] In discussing the symbolic nature of physical theory, Cassirer makes this same point with regard to causality:

It contains in its meaning the claim of "always and universally," which experience as such is never warranted in making. Instead of deriving a principle directly from experience we use it as a criterion of experience. Principles constitute the fixed points of the compass that are required for successful orientation in the world of phenomena. They are not so much assertions about empirical facts as maxims by which we interpret these facts in order to bring them together into a complete and coherent whole.[81]

The philosophical objections to quantum theory seem therefore to be of two kinds, both aimed at its claims to be a causal or deterministic theory of the traditional kind expected in Newtonian and Einsteinian terms. The first, presented by Einstein himself, as well as Schrödinger, Bohm, and DeBroglie, was concerned with the probabilistic form of the theory. Einstein stated that quantum theory did not supply a complete enough description of the state of objects as they moved through space and time and thus of physical reality as it exists external to man's perceptions.

The second objection, by Nernst, Heisenberg, and Northrop, accepts the probabilistic form, separates it logically from traditional statistical theories and reads into the difference in the kinds of causal relationships some novel or mysterious quality which has been introduced into reality.

As pointed out above, Cassirer would reject the first or formalistic demand as being too restrictive of the structure of scientific theory and ultimately leading to the same errors underlined in Kant's commitment to Newtonian apriorism. The error derives in short from an overemphasis on physical realism in scientific philosopy. The second objection is also derived from a psychological demand which man finds hard to resist: his ascription of a content to reality which supposedly resists or challenges man's descriptive and theoretical efforts. To describe this content one has to indulge in a form of metaphysical realism which necessitates a language using words such as "ontological potentialities," "teleological causations," "purpose," "free will," which carry us far beyond the contextual circumstances of the experimental materials.

It is an attempt to transcribe what is in fact a methodological limit of scientific enquiry into an ontologically significant fact.

Cassirer attempted to counter claims that meant to impute metaphysical implications to the uncertainty relations: "They [uncertainty relations] are not categorical statements about the objectively real, but rather modal statements about the empirically possible, about the physically observable. They accordingly do not presuppose a definite object only to determine subsequently that it will never be entirely accessible to our knowledge; rather they contain a new stipulation concerning objects, which we may rightly use as long as we adhere strictly to the limits of the observable."[82]

In replying to the realists, Cassirer proposed that the answers given to us by nature are determined not alone by nature but also by our particular instruments and the kind of observations we engage in.[83] It is thus impossible to draw a sharp line between nature and our knowledge of nature, "the two are inseparably interconnected." "Nature" has no characteristics which have not in some way been subsumed to science. Further, we have no right to assert that nature contains in itself a measure of inscrutability which sets permanent limits on the capacity of science to penetrate the knowable unknown.[84]

The following somewhat enigmatic statement epitomizes the issue: "Then nothing that is not 'for us,' that is not for physical knowledge in any sense, is any longer in 'itself' in nature. The type of determination prescribes limits to the being which we can attribute to natural things, and not the reverse. It is not being, determined in itself, that sets permanent limits to knowledge, with the absolute intrinsic nature of being remaining impenetrable."[85] In other words, natural things do not prescribe limits to what can be attributed to them. The type of theoretical determination we make prescribes the limits of what can be attributed to the nature of things.

It may at first appear as if Cassirer is making a rather unphilosophical crusade against interpretation in physical theory. It is as if he is telling the scientist to search for any kind of law, so long as it (a) fulfills the truth function requirements of scientific method, and (b) that it will propel science toward (1) unifying the proliferation of experimental data in ever more encompassing theoretical systematizations, and (2) freeing this trend toward abstract and relational envisagement from pictorialism with its objects and models, e.g., those qualities of thought which bind man to the necessities of common-sense experience. Nevertheless in Cassirer's neo-Kantian idealism there is a philosophical perspective and direction which, while it does not intersect his study of the status and structure of scientific theory, does give science philosophical significance beyond purely epistemological, experimental, and methodological concerns. This philosophical dimension transcends the metaphysical aspirations of many scientists in this discipline.

Thus our analysis of the structure of physical theory thrusts us back again from the external and objective reality of things to the problem of the structure of thought. There is the question then, does physical theory tell us more about man, about the nature of his thought than it does about external reality? Perhaps the question is unanswerable, since our theories are couched in terms of the symbols describing the invariables of external experience. Whether the two worlds, the outer and physical world and the inner and psychological world, can be joined in some common language is a question akin to those questions derived from quantum theory concerning free will and indeterminacy in social experience. They go beyond the bounds of the symbols of the theory itself. While we cannot rule them out of order as philosophical issues, they necessarily inhabit the world of imaginative thought and of possibility.

The status of these conjectures is similar to that of such general terms as "external reality" or "things in themselves." They are in the Kantian sense "regulative ideas" in that they do not limit thought to the strict bounds of epistemological validity but serve, in marking the boundaries between knowledge and hypothetical possibility, to direct the imagination into new patterns of thought.

By attempting to shear away traditional metaphysical ascriptions of significance to physical science, Cassirer did not propose to abandon the attempt to understand science philosophically. He meant us to withdraw from a preoccupation with physical or psychological reality and see science in a larger context of human thought and action. The philosophical significance of science would then be sought in its function for thought in the richer experiential dimensions of culture and in the diverse symbolic systems within which social man is observed.

> The physical concept of reality should ultimately be so formulated as to unite the totality of aspects resulting for different observers, so as to explain them and make them understandable; but in precisely this totality the particularity of the viewpoint is not extinguished but preserved and transcended. In this whole movement scientific knowledge confirms and fulfills, in its own sphere, a universal structural law of the human spirit. The more it concentrates in itself, the more clearly it grasps its own nature and strivings, the more evident becomes the factor in which it differs from all other forms of world understanding, and the meaning which links it with them all.[86]

5

From Science to Culture: *Language*

LET US SUMMARIZE THE MAJOR POINTS OF THE argument thus far. When one examines the wide range of philosophical interpretations given to science, from Galilean dynamics to quantum mechanics, it becomes apparent that the study merely of the laws which guide reason in stabilizing the flux of phenomena is not in itself enough. The ordering of material objects into lawlike regularities and the ability to predict future events lead the enquiring mind to search for the meaning of these scientific phenomena.

Thus we have seen that from Galileo to Hume the problem of scientific and then deterministic knowledge was set within a context in which the problem was always broadened to include metaphysical questions as to the nature of reality: whether it was indeed mathematical in structure; whether simple ideas could partake of this essential quality of the real; whether knowledge could be reduced to certain psychological states in an individual's mind; or perhaps could the entire process of scientific knowing be a particularly human aspect of a larger content of the divine. Man could not rest merely with a knowledge of the contextual functions of science. The philosophical drive to understand significances and relationships was ever dominant.

Even in the Kantian syntheses, which shunned metaphysics and labeled these conjectures on the nature of noumena as regulative ideas which stir the human imagination but cannot partake of the same logical characterizations as those constituting knowledge, there is a search for significance. But philosophical enquiry as to what constitutes the context of scientific knowledge is now broadened in Kant to a phenomenological search into the conditions of knowing in the ethical, esthetic, and biological spheres. If we are not to be introduced into the mysteries of psychology or metaphysics through the power of scientific knowing, then scientific knowing must be silhouetted by a wider investigation into the development of a variety of rational principles, each with its presumably *a priori* forms of thought which order experience. Thus, as Martin Buber has hypothesized, Kant brought us to the door of philosophical anthropology, but unfortunately, although the questions loomed large for him, he never stepped fully into that chamber of mysteries implied by the problem: what is human nature?[1]

In the generations which followed the death of Kant, the Hegelian search for a metaphysical understanding of reality behind the facade of the phenomenal interpretation of experience then enveloping philosophy allowed science to free itself temporarily from philosophy and pursue its own inner development. But as we have seen, the growth of Newtonian physics and the power of this science to reveal lawful regularities within a constantly growing range of experience within "physical reality" again seemed to demand interpretations. There were few scientists who could surrender their search for explanations and say with Hertz that the meaning of Maxwell's equations resides only in Maxwell's equations. A great dialogue between scientists from a number of different disciplines, each with its diverse claims for priority, and from a number of different philosophical positions was necessary before it was finally realized that, logically speaking, the structure of scientific laws could not be utilized as evidence for any claims in the attempted description of ultimate reality. Mechanics (Newtonian physics) was not the

form of the physical universe nor was energetics or thermodynamics. Physical reality, likewise, could not be reduced to a
series of phenomenally neutral entities—because of their psychologically prior and sensationally uncorrupted state in the human
consciousness (Mach).

Einstein's theory of relativity underlined the independence of
the physical interpretation of space, time, and motion from any
special and primary referential system. It was essentially a de-
ontologizing theory, however, which in its absorption and reinterpretation of the Newtonian theory fulfilled and completed the
classical logical scheme for the nature of scientific objectivity.
The challenge of quantum theory, with its radical probabilistic
revision of traditional mechanical determinism, stirred into being
once more a number of interpretations of physical theory which
themselves transcended the bounds of the theory itself.

The metaphysical ascriptions given to relativity theory, as for
example by theologians or moralists who saw in it the disintegration of traditional absolute truths and standards, needed much
less clarification on this issue than quantum theory. Now, the
very fact that traditional "state" descriptions of the objects of
physical science were altered caused great concern to many
scientists, including Einstein, who though very clear in his
neo-Kantian instrumentalism still made certain specific logical
demands of physical theory, as absolute in their own way as
those of nineteenth-century physicists. These Einstein could not
abandon. So too, Heisenberg, Northrop, and other thinkers found
it difficult not to go beyond the structure of quantum theory
to postulate wider philosophical, or better, ontological, significance to this departure from the traditional forms of our causal
principles.

Cassirer held fast to a neo-Kantian instrumentalism which
demanded no special form for scientific theories, except that they
be freely constructed without dependence on any given model,
whether common-sense or mechanical, and that they meet the
varying contextual demands, experimental and conceptual, of
science. In addition he held that we ought to expect that these

laws would tend toward ever wider and more universal relational structures of thought.[2] As long as theories met these logical canons, as long as they were fruitful in guiding experience through the predications of thought and experiment, they would fulfill their obligations to the scientific tradition.

Certainly the evidence for this historical trend in scientific theory and the increasing number of scientists who accept instrumentalist interpretations of science support Cassirer's analysis. The question then arises, why is this trend evident? What philosophical significance does it have? Cassirer never faced these questions directly. While he did not engage in a polemic against metaphysical explanations in science, as did Pierre Duhem, he was certainly just as suspicious of any attempts to propose deeper explanations. Instead, Cassirer preferred to illustrate these trends by referring to their historical context, interpreting his method of enquiry as a phenomenology of thought. His subsequent approach was to illustrate the same movements of thought as they took place in other disciplines.

The question as to the meaning of the evolution of science, whether it was approaching truth, objectivity, or ultimacy, bordered too closely on metaphysics to attract Cassirer. He was equally evasive on the question of how one might judge the proper conceptual or practical utility of a theory in order to choose between alternatives. The reasons for choosing theories could only be derived from considerations outside of science. Cassirer gave rich and copious evidence to illustrate how men had decided and the criteria they used for their decisions. His evidence came from an enormous range of concrete theoretical instances in all fields. The only message which was clear was that absolute objectivity, ultimacy, and permanence as conceptual components of truth were to be avoided.

An interesting illustration of this issue is given by the manner in which such discussions have evolved. In the nineteenth century, scientists such as Ernst Mach, Karl Pearson, John Frederick Herschel, and William Whewell spoke of science as "economy of thought" or "descriptive shorthand." These aphorisms were an attempt to describe the functional use of theory in its practical confrontation with experimental materials. It was also an attempt to prescribe for other scientists what was to be the form or goal of theory making.[3]

Philipp Frank, although taken with the positivistic purity of these precepts, added a new dimension to the criteria scientists apply to theory to give direction to the evolution of science. He first stated, "there could be a theory which agrees with all observed facts but is a mere record of observations and no theory at all. If we have two theories which yield the same observable facts, the scientist prefers the theory which is more economical or just simple."[4] But this is still not an explanation as to what criteria one will invoke in order to choose the simple theory or to justify the criteria. He develops this problem in a later passage, in which he states that when all predictive and experimental factors are neutralized the scientist chooses the simplest or most beautiful of the theories.[5] Thus functional, practical, or conceptual shorthand as a criterion for our acceptance of theory and our guide as to what to look for in future theories is now transformed into an esthetic criterion. We are catapulted into a wholly new and unusual problem of justification and explanation in the history of scientific thought. That this characterization of the scientific enterprise in terms of esthetic values is not an unusual one is exemplified in a most surprising statement attributed to the scientist-philosopher, Hermann Weyl: "My work always tried to unite the true with the beautiful, but when I had to choose one or other, I usually chose the beautiful."[6]

1

Throughout his career as a student of the philosophy of science, Cassirer fought against substances, pictures, sensations, and all other reductions that metaphysically complicated science. Yet in keeping science epistemologically pure, so to speak, Cassirer was not trying to isolate science intellectually. On the contrary, as early as *Substance and Function* (1910), and surely by the time he wrote *Einstein's Theory of Relativity* (1921), two powerful influences were united in his thinking to guide his functionalist position in science and to provide the psychological support to withstand the lures of metaphysics as he assayed the philosophical significance of science. The influences were both historical and contemporary. They were historical in the sense that, as a Kantian, Cassirer was aware of the direction of Kant's philosophic efforts. Kant never intended the *Critique of Pure Reason* to be an end in itself. It was a beginning, in that by drawing the epistemological lines for discursive knowledge, Kant would be able to go on to the (to him) more crucial issue of ethics, as well as esthetic and biological thought. He had almost to create these intellectual disciplines in terms of the questions one must ask of them in order that they be subject to philosophical analysis.

As for contemporary influences, there existed for Cassirer a well-defined, indeed a rich *geisteswissenschaftlichen* empirical and theoretical domain. Cassirer's problem was not unique to his time, that of bridging the realms of knowledge in the context of a "two culture" intellectual environment in which science as a pillar of rationalism was opposed to the other "nonrational" disciplines. On the surface Cassirer's classically instrumentalist view of scientific theorizing would serve only to isolate science increasingly from philosophy. This was not his intention. For if the result was to undermine traditional philosophical claims for science, it meant only that rather than despair over past philosophical failures, a newer and more firmly planted bridge was needed. It was toward this end that the greater part of his subsequent philosophical efforts was directed.

Cassirer's Kantianism stood him in good stead, for while the content of the various disciplines was quite different than in Kant's time the program was somewhat similar. There was of course no intention of using scientific knowledge to "liberate" man in another cultural realm. Cassirer had no preconceived notion as to the results of his enquiry. If there was a theme in the back of his mind, it was that the high valuation so many intellectuals placed on science (derived in part from its tangible predictive power) did not necessarily imply the derogation of the less strictly deterministic cultural disciplines, in spite of the historic prejudices left behind by the metaphysical absolutism of Hegel and his disciples.

The humanistic forms of thought were now of crucial importance to Cassirer's understanding of science. In a sense they became a mirror for the further significance of science. For there existed no ultimate scientific "reals" which were ontologically secure and independent. We could no longer point to a reality which science made evident to thought. Nor was there a reality outside of our known scientific symbols which science attempted to uncover but which was ever thwarted by intrinsic limitations within its structure.

The meaning of scientific thought then could be found by examining the context in which the scientific endeavor was but an expression. Science was an instrumentality of thought used to order experience. But there were other forms in which thought found expression and other aspects of experience which were revealed and cognized in their own particular manner. The role of the philosopher was not only to disclose the unique character of these differing "tools" of thought—the particular symbolic envisagement given to experience by their special modalities— but also to ascertain the connection between these domains of knowledge. Thought is whole. And if science was to be explained in the context of being one particular among others, thought had to be seen as one in spite and perhaps because of its diverse manifestations. In the tradition of Kant, Cassirer had to create for himself a philosophical anthropology through which this program would be carried out.

The problem of the context of science, then, necessarily leads to the question of the law or principle of connection. Just as one could not justify the abandonment of the study of science for the humanities and, in this, hope to form a more complete picture of thought, so too the mere placing of science among the larger number of cultural disciplines would be insufficient. The differences and similarities between the varying dimensions of knowledge must be sought in the nature of each discipline—in its logical principles of structure, of formation, and even in its patterns of evolutionary development. For example, the character of the physical sciences is defined as a process of transforming subjective and qualitative factors into purely objective mathematical concepts: "it is precisely its specific cognitive task to translate everything enumerable into pure number, all quality into quantity, all particular forms into a universal order and it only 'conceives' them scientifically by virtue of this transformation."[7]

But this does not complete the philosophical task, for just as the concepts of causality or of space and time have a different meaning and form in mathematical physics than they do in common-sense discourse, so are there other areas of thought in which space and time are given form, structure, and meaning still different from that found in the above two areas.

> The moment that we transcend the field of physics and change not the means but the very goal of knowledge, all particular concepts assume a new aspect and form. Each of these concepts means something different, depending on the general "modality" of consciousness and knowledge with which it is connected and from which it is considered. Myth and scientific knowledge, the logical and the aesthetic consciousness, are examples of such diverse modalities.[8]

This is true also with regard to space and time. The historian views time quite differently than does a mathematician or physicist. Painting and architecture develop particularly unique spatial concerns. Further, the manner in which music was united with mathematics, under the Pythagoreans, represents a special synthesis subsumable to neither discipline. So too in music, rhythm and time are developed for different purposes than the rhythms and temporal sequences of action in biological forms.[9]

When in 1920 Cassirer moved to the University of Hamburg, he found waiting for his use the library of the Warburg Institute, where a wealth of material had been gathered in the fields of linguistics, ethnology, comparative religion, and anthropology. The resultant shift in intellectual emphasis resulted in the writing of *The Philosophy of Symbolic Forms,* from which Cassirer's more mature philosophical reputation derives. It is a shift from the perspective of the physical sciences alone to the larger world of culture, of language, myth, and religion, art, and common sense, all steeped in Cassirer's traditionally rich evocation of the history of intellectual thought. But while the perspective is somewhat altered, the concern for an understanding of scientific reason remains.

There is in this shift in perspectives an implicit assumption that the evolving pattern of theory making in the physical sciences is not in itself a unique historical event. We have witnessed the gradual emancipation of physical theory from the qualitative substantialism first manifested in Aristotelian and scholastic thought and the freeing of modern science from the intellectual grip of recent materialistic and sensationalistic reductionisms. This has finally led to the symbolic and abstract mathematical envisionment of physical experience as being part of a more general intellectual process.

Cassirer approached the cultural disciplines with the Kantian expectation that these diverse expressions of thought would also display similar patterns of structure both in their historical and systematic development. There were two basic criteria derived from his study of the physical sciences that he felt needed confirmation in the cultural disciplines to prove his assumptions concerning the inner symbolic trend of conceptual thought. These were (a) the test of ideality—the exhibiting of a movement away from images and thing concepts toward ever more abstract symbol systems and (b) the encompassing by symbols of more generic or universal meanings. This trend ought to disclose a direction leading from the specific to the general, incorporating ever more intellectually economical structural solutions to perceptual experience.

These were quite general theoretical expectations that would act to a great extent as a heuristic guide to further research. Cassirer naturally expected that the very obvious differences in the external form and function of other patterns of thought would also have their unique theoretical correlates. The patterns by which physics developed its abstract mathematical as well as universal theories would not necessarily carry over to other disciplines.

Thus in his studies of mythic and religious symbols and to a lesser extent in his study of esthetic symbolism, he did find trends in the development of cultural expression toward the creation of ideal and formal elements, e.g., in religion, monotheistic thought, as a late development as compared with "thing gods" or "place-name deities" found in earlier strata of culture.

But the differences in the referential scope between religious symbols and scientific symbols were just as important as their similarities. Scientific laws are always presented in reference to a larger spectrum of relations. The particular concept in science receives its signification by reference to its logical place in the general system of scientific laws in nature. The vector from the particular concept leads outward to more universal principles and the principles lead the mind back toward the particular phenomena which often become the predictive or the logically deductive consequences of the theoretical principle.

In mythic and religious symbols, as well as those of the esthetic realm, meaning inheres in and is exhausted by the individual symbolic experience.[10] In religion and myth the experience is personal and emotional and refers to the qualitative feeling-centered aspects of the person. In art the experience is sensuous and direct and likewise refers to and is exhausted in the perceptive moment. Neither of these symbolic forms refers to a wider nexus of meanings in order to be understood. In addition the discursive and denotative element is an ancillary aspect of the nondiscursive cultural experience in that we may speak about religion or art in a descriptive or critical manner, but, to use a Deweyan term, the "consummatory" aspect of such symbolic experiences refers only to the individual's emotions and perceptions and does not gain intrinsic significance from scientific facts and theories.[11]

2

Cassirer's enquiry into the significance of scientific thought received an important impetus when he turned to the study of language. He began this study by examining the historical tradition of linguistic theory as it developed in recent times. He immersed himself in the vast accumulation of concrete empirical research that had been gathered in the field and was available for his use at the Warburg Institute.

Three concepts—space, time, and number—were used to reflect varying dimensions of linguistic development and structure. The comparative data which Cassirer gathered seemed to confirm his expectation that even in the seemingly casual assigning of meaning to sound in common-sense language a pervasive logical trend could be discovered. This trend, beginning with the use of language in terms of the immediacy and discreteness of sensory experience, evolved toward the building of more relational and abstract syntactic and grammatical forms.

Cassirer employed space, time, and number because he hypothesized that for common-sense thought these concepts embodied a universal set of experiences which would be immediately reflected in the language of the respective tribes and societies. Although the conceptual articulation of these constructs in the various systems of meaning was not synchronous in all cultures (time, for example, being far more diffuse and abstract a concept, and therefore linguistically more amorphous and undeveloped than space, number being even more abstract than time), he found the characteristic logical evolution of these concepts to be similar.

Thus in the earliest phases of language—what Cassirer called the mimetic stage—language reflects an extraordinary attention to individual nuances of perception and experience. "Here language clings to the concrete phenomenon and its sensory image, attempting as it were to exhaust it in sound; it does not content itself with general designations but accompanies every particular nuance of the phenomenon with a particular phonetic nuance, devised especially for this case."[12] Cassirer cites Hammer-Purgstall, who noted the Arabic use of between five to six thousand terms to describe a camel, none expressing a general biological concept.[13] Karl von den Steinen reported that the Bakairi of Brazil use a language that is literally choked in an abundance of particular notions attached to particular things and manifests, as illustrated in their vocabulary, little interest in common characteristics.[14] One might add the report of Jespersen, who notes that many languages had not in historical times attained the use of generic terms for certain colors.[15] Instead the colors are assigned linguistically to the object spoken about, e.g., canary yellow, blood red. Cassirer noted, "Particularly the languages of primitive peoples are distinguished by the precision, the almost mimetic immediacy, with which they express all spatial specifications and distinctions of processes and activities. The languages of the American Indians, for example, seldom have a general term for 'going,' but instead possess special terms for 'going up' and 'going down' and for countless other shadings of motions; and states of rest—position, standing below or above, inside and outside a certain limit, standing near something, standing in water, in the woods, etc.—are similarly differentiated."[16]

This primary intellectual striving by man to absorb and stabilize the plethora of sensory perceptions into language is an inherently chimerical enterprise. The malleable qualities and the relative discreteness of sounds lead to a gradual loosening of the bond of sound and sense so that a more general and formal analogy can be established between the phonetic sequence and the meaning content.[17] This next stage is called the analogical, in which, as in Indo-Chinese, Sudanese, Ethiopian, and even Finnish, tonal gradations are used to indicate differences in meaning. Here the relationship is quite conventional in that high or low sounds do not have a fixed sensory signification. Yet the fact that meaning is shaded through a sensuous rather than through a structural alteration of the word or sentence connoted to Cassirer a simpler cognitive stage in the development of the linguistic formalism. Thus as language gradually frees itself from the sensory, the concrete, even the individual centeredness of experience, it opens up for itself potentially wider worlds of meaning.[18]

The individual begins to see space, time, and number as abstract notions. The immediacy of things is left behind as language reveals new levels of subjectivity and objectivity, of action and volition, of conditional and assertive modes of thought. As these nuances are articulated in consciousness, language itself develops a structure in which, while the parts of speech are differentiated from an ever fluid condition of shifting meanings, the one-word sentence (*e.g.,* "tl 'imshya 'isita 'itlma 'nootka"), a paradigmatic manifestation of this stage, gives way to a grammatical solidity of structure which then acts as a formal expression of a cultural community's sense of experience and continuity.[19]

This is not to imply that the mind is passive to the dynamics of language. For there is an interaction between the inner rational drives of thought and the existent cultural formulations of thought: " . . . we do not simply seize on and name certain distinctions that are somewhere present in feeling or intuition; on the contrary, on our own initiative we draw certain dividing lines, effect certain separations and connections, by virtue of which distinct individual configurations emerge from the uniform flux of consciousness."[20]

The tension that exists between the inner movement of thought as it alters the language in each generation and the existent state of linguistic expression within the culture can be noted from another point of view. Common-sense language at the stage of development we are now discussing is distinct from purely logical or scientific modes of thought which developed later in the evolution of culture. The differences in the use of language which distinguish the development of a higher civilization do not depend on a new linguistic structure or vocabulary. By merely noting the social forms of communication in terms of grammar and vocabulary, it would be difficult to distinguish early civilizations such as Egypt or Sumer from Hellenic Greece or even the modern world. The commonalities of ordinary existence have been universal for some five thousand years. However by noting the attitudes and intentions which language serves, one realizes that a crucial shift to a logical and conceptual awareness of the use of words has effected a great transformation of language. This transformation is of the same order as that which attended the transition to ordinary common-sense modes from the earliest proto-mythological uses of language.

This relation of language and logic is posited by Cassirer as a distinction between logical and linguistic concept formation. Linguistic concept formation "never rests solely on the static representation and comparison of contents but that in it the sheer form of reflection is always infused with specific *dynamic* factors; that its essential impulsions are not taken solely from the world of being but are always drawn at the same time from the world of action. All linguistic concepts remain in the zone between action and reflection. Here there is no mere classification and ordering of intuitions according to specific objective characteristics; even where there is such classification, an active interest in the world and its constitution expresses itself."[21]

This is but the first phase of man's distillation of perceptions into those concrete intuitions of thought which endow perception with a representational ordering of the familiar relations of I-thou-he and the various modalities of the spatio-temporal order.

If we conceive of the whole intuitive world as a uniform plane, from which certain individual figures are singled out and differentiated from their surroundings by the act of appellation, this process of specification at first affects only a particular, narrowly limited portion of the plane. Nevertheless, since all these individual areas are adjacent to one another, the whole plane can gradually be apprehended in this way and covered by an even denser network of appellations. Yet fine as the meshes of this net may be, it still presents gaps. For each word has only its own relatively limited radius of action, beyond which its force does not extend. Language still lacks the means of combining several different spheres of signification into a new linguistic whole designated by a unitary form. The power of configuration and differentiation inherent in each single word begins to operate, but soon exhausts itself, and then a new sphere of intuition must be opened up by a new and independent impulsion. The summation of all these different impulses, each of which operates alone and independently, can form collective, but not truly generic unities. The totality of linguistic expression here attained is only an aggregate but not an articulated system; the power of articulation has exhausted itself in the individual appellation and is not adequate to the formation of comprehensive units.[22]

But the mind cannot rest here. The human being even in his most prosaic activities wants to give structure and organization to his experience. There is thus a precipitous moment when language takes a further step toward generic universality. Whereas it had previously created specific words for specific intuitions, now such verbal designations are combined to reflect the more objective relations perceived by the user.

Thus according to Cassirer the transition from purely qualifying concepts to classifying ones takes place as a reflection of the attempt by thought to create a stricter relationship between sound and meaning by coordinating a specific series of conceptual significations with specific phonetic senses. The simplest example is that of marking groups of words with a common prefix or suffix. The Indo-European words for family relationship—father, mother, brother, sister, and daughter, etc.—are examples of this form of concept formation. The endings in their original form (-tar [ter], as in *mātár*, *bhrátar*, *svásar*, *duhitár*, etc.) are meaningless outside the context of these words, and provide the conceptual link between thought and linguistic expression which begins to build the intellectual perspective of a language family.[23]

If this process should not be conceived as a logical process in the strict sense, Cassirer likewise argued against Wundt's opinion that this linguistic concept is developed through associative processes. As Wundt states: "For it is evidently the expression of affinity immediately presenting itself in the transition from one object to another that constitutes the *concept* of affinity, so that this concept rests rather upon certain similarly colored attendant feelings than on actual comparison."[24] Although the motive for the comparison rests upon feelings, the very fact of specification in language Cassirer believed to be an "independent logical act with a specific logical form. A determination which remained exclusively in the sphere of feeling, could not by itself create a new objective specification. . . . Feeling can join anything with anything; hence it provides no adequate explanation for the grouping of *specific* contents into *specific* unities. For this we rather require a logical basis of comparison, and such a basis is discernible in linguistic series even where it is expressed only in the form of a classifying suffix and not as an independent substantive."[25]

The structure of language continues to be differentiated in this manner, depending thus far upon the particular and unsystematic interests, outlook, as well as other psychologically relevant factors in the experience of the society. But it is important to note that this growth of conceptual organization does not necessarily reflect contemporary scientific or logical criteria of thinking. Thus the Melanesian and American Indian languages employ special prefixes for round or long objects. The relational family consists of an odd assortment of round objects—moon, sun, certain fishes, the human ear, etc. On the other hand, the linguistic classification of a group of long objects brings together the nose and tongue.[26] In languages such as the Bantu, relations rather than contents form the basis for linguistic differentiation, as for example certain prefixes for large objects and others for small ones. Some American Indian languages, further, classify words as to whether one thinks of something as standing up or lying down. "Such classifications make it very clear that the first conceptual differentiations of language are still thoroughly bound up with a material substrate, that the relation between the members of a class can only be *thought* if it is *embodied* in an image."[27]

Cassirer notes the slow infusion into all language structures of more imaginatively based relational elements. Such students of language as Brugmann and Grimm gradually came to realize that the common forms of gender classification were part of universal trends toward generic classifications.[28] But the tendency to go beyond immediate consciousness of sex differences to a creative imputation of gender shadings in language was interpreted by Cassirer as being only one aspect of a general trend in an overall human need to order experience. Humboldt, in his studies of the American Indian languages, found what he thought was a deeper stratum of thought, the difference between animate and inanimate.[29] Another example is Franz Müller's discovery that, in the Dravidian languages, all nouns are divided into a class of "reasoning," which belonged to man, gods, and demigods, and "unreasoning," which belonged to animals and inanimate objects. These provided further evidence of a powerful trend toward classification in all forms of experience—observation, mythical and imaginative thought, and language formation.[30]

Language, Cassirer claimed, never limply follows the random lead of impressions and perceptions, but orders them in a dynamic manner. Thought, by distinguishing between impressions, choosing and directing them through language, created certain objective centers of interest. At this stage however the world of sensory impressions is permeated not only by a proto-judgmental capacity and "the theoretical nuances of signification," but with emotive valuational nuances. These two elements continue to interact, to "shade off continuously into each other." But language reveals the inner logic from which it is being constructed. The first structural distinctions do not vanish, but tend to endure; not only do they persist in their logical necessity and consistency, but they begin to affect other areas of language formation with like logical perspectives.

Through the rules of *congruence* which govern the grammatical structure of language and which are most clearly developed in the prefix and class languages, the conceptual distinctions applied to the noun spread to the other parts of speech. In Bantu every numeral, adjective, or pronoun which enters into an attributive or predicative relation with a substantive must assume the characteristic class prefix of that word. Similarly, the verb is allied with its subject and object by a special prefix. Thus, the principle of classification, once arrived at, not only governs the formation of nouns, but thence spreads to the whole syntactical structure of the language, becoming the actual expression of its organization, its spiritual "articulation." Here the work of the *linguistic imagination* seems throughout to be closely bound up with a specific methodology of linguistic thinking. Once again language, with all its involvement in the sensuous, imaginative world, reveals a tendency towards the logical and universal, through which it progressively liberates itself and attains to a purer and more independent spirituality of form.[31]

3

Cassirer cites the ultimate development of certain distinctive groups of languages as a concrete but universal manifestation of this logical and ideational process which operates within the raw sensory materiality of sound and which transforms this undifferentiated sensuous residuum into a structure of meaning. These language orders each reflect a systematic resolution in linguistic form of a universal logical problem. In each specific case a similar logical result has been brought about through and in spite of the particular historic conditions which created the language family. Thus, for example, at one end of the linguistic spectrum lie the isolating, analytic types of languages, such as Chinese, in which meaning is determined by the manner in which individual words are ordered in the sentence. The arrangement of the discrete fragments is crucial here to the meaning.

Midway in this arrangement of the varying language structures are those grammatically inflected and synthetic languages represented respectively by the early Indo-European tongues, such as Latin and Greek, and the later Anglo-Saxon and Germanic forms. In neither of these types are the order and location of the elements crucial to the meaning of the sentence, whereas modern languages have become more analytic or isolating in their grammatical structure. The other extreme is represented by those American Indian languages in which one word often provides the meaning for an entire sentential idea.[32]

Cassirer was concerned that no innate superiority or inferiority be imputed to any one of these linguistic logical schemes. His position has gained support among contemporary anthropologists and linguists.[33] It is difficult to decide whether or not one spoken language is more suited than another for the expression of the higher level abstractions necessary for civilized life. In certain societies there are so many cultural and historic reasons for the existence of written language, which is surely the *sine qua non* for advanced forms of civilized life, that it is impossible to attribute the causal factor to the spoken language *per se*. Certainly after the fact it seems obvious that Greek is the appropriate language of philosophy. But it might just as well have happened that other primitive languages, given an exposure of its society to life in the Mediterranean in the first millenium B.C., would have developed similar philosophies from a different linguistic base.

Using Cassirer's schematism we may ask whether spoken language still in the mimetic or analogical stage could with the addition of a twentieth-century vocabulary make an easy and suitable adaptation to modern world conditions. We might note here that Finnish, through the use of a written script has adapted itself to modern intellectual needs and that in general, as both Kroeber and Carroll have pointed out, primitive languages which classify by shape or have widely developed active or passive modes could well relegate these grammatical patterns to a purely formal status (thus to simply ignore them) as has been done with German and Russian with regard to their purely formal gender classifications. The semantics of these languages could then be reintegrated in a functional and instrumental manner to meet new cultural needs without necessarily destroying the traditional grammatical structure.[34]

Cassirer is not subscribing here to a Whorfian view of language: namely, that each language acts as an intellectual filter for experience or that language, in a literally idealistic manner, creates a world view for man out of which he can never break and which in fact is incommensurable with the picture of experience which other languages present.

Cassirer recognized that language is the product of, and in turn conditions a host of, subtle ritualistic, interpersonal, and esthetic cultural nuances which are untranslatable from one language to another. This problem of course exists in the literature of even closely related languages coming from similar cultural backgrounds. Here we deal with nondiscursive symbolic elements which constitute the other logical side of the linguistic coin—elements which are involved, as is the spoken dimension of language, with an ultimately dynamic, emotive innovative dimension of the language problem. It is true, nevertheless, that the universal exigencies of changing environmental and ecological conditions necessitate that all people must somehow transcend their intuitive linguistic patterns of experience and act in accordance with the necessities of life. One can thus hypothesize that both the inner personally innovative and the external environmental dynamics of life act so as to enforce a continuing alteration of language content and structure.

Cassirer's concern was to note that within this dynamic there exists law or an inner trend, intuitively expressed, that is manifested in the envisioning of experience in always more abstract and relational categories and in always less concrete and perceptual terms.[35] In Sapir's phrase, modern languages tend toward structural simplicity. This logical trend is seen in the systematic character of all language orders—whatever their type and whatever their particular manner of development—in which one organizational principle will dominate and become the logical center. This pattern will then be radiated to the extremities of the system and will condition the subsequent evolution of the language family. Edward Sapir expresses this intuitive logical trend of thought as follows:

> The enormous amount of study that has been lavished on the history of particular languages and groups of languages shows very clearly that the most powerful differentiating factors are not outside influences, as ordinarily understood, but rather the very slow but powerful unconscious changes in certain directions which seem to

be implicit in the phonemic systems and morphologies of the languages themselves. These "drifts" are powerfully conditioned by unconscious formal feelings and are made necessary by the inability of human beings to actualize ideal patterns in a permanently set fashion.[36]

It is the development of written language which marks the transition from the intuitive linguistic world view to a more discursive and cognitive representation of experience. This in turn leads eventually to mathematics and science and the envisioning of experience as permanent, quantitative, and universal. As Kroeber notes, "This typical continuity of change is what characterizes spoken language as being natural, spontaneous, underlying growths, in contrast with their superimposed and more 'artificial' writing systems, which quickly tend to crystallize. The difficulty with any *spoken* language is to keep it the same; with its *written* expression, it is to keep it plastic. The writing will set and finally break in a sort of revolution, but it has great difficulty in conforming and adapting. Written languages that have survived for millenia, like Latin and Sanskrit, are dead as spoken languages."[37]

Both of these languages—like the Hebrew now revived and modernized in Israel, and Arabic, with its numerous "provincial dialects," now approaching a point of formal separation from the classical script—represent the linguistic consequences of the psychic impact on history of great cultural traditions. Chinese linguistic evolution also manifests certain similar trends. Their written system is ideographic or "logographic," organized around word units, rather than being a transcription of phonetic sounds. China, subsequent to the great period of classical literature, tended toward greater cultural diversity, which in time gave rise to relatively independent spoken dialects of the same written language that are now mutually unintelligible in North and South China.

Kroeber notes that the written language of the Chinese during China's great period of cultural advance was taken over by the Annamese, Koreans, and Japanese, but with differing pronunciation of the original characters in each case. Thus in Japanese there are certain ideographs which have completely different-sounding pronunciations from the original Chinese. Along with these, purely Japanese sounds have been attached to the same symbol. In both cases we have a similar meaning, but with entirely different sound realizations of the original written character.[38]

One can hypothesize then, that endowing sensory experience with signification through sound was a crucial step in man's realization of his power to theorize. It is sure that he did not, in the earliest eras, understand the import of this spontaneous act, an act which constitutes the most obvious and practical realization of a conceptual need that is rooted in the structure of human thought.[39]

Speech points to broader issues than oral language itself. For, when we arrive at the point of creating written and therefore permanent, ideal and general symbols of thought, the ground is prepared for a new logical and theoretical advance by man. As Cassirer pointed out, this comes about when man is able to step back and examine the presumptive meanings of words, as when Socrates criticized the ordinary vague and contradictory common-sense usage of language in his day in an attempt to establish a logical and consistent usage. "And indeed, historically speaking, the problem of the concept was discovered when men learned not to accept the *linguistic* expression of concepts as definitive, but to interpret them as *logical questions*."[40] And with the birth of systematic philosophy the ground was prepared for the coming of science.

6

Aphasia and Common-Sense Thought

BEFORE KANT, THE SEVENTEENTH- AND EIGHTEENTH-
century philosophers had sought to find meaning and significance
in science by trying to identify it with a reality that was substantial, generally external to man, preexistent, and permanent. In
other words the significance of theory lay in its ability to reveal
a meaning which transcended science itself. In the nineteenth
century scientists, foregoing the pretentions of ontology in general, concerned themselves with the meaning of theory, generally
Newtonian mechanics. These laws and theories—the symbolic
mortar of knowledge—were identified with the real and the permanent. There thus existed for man a fixed structure of external
theory. The meaning ascribed to science was not as far-reaching
metaphysically as in the earlier era, though in its own way it was
certainly interpenetrated with metaphysical and realistic status.
These nineteenth-century presumptions were effectively punctured by the symbolic theories of knowledge.

The substantiation of the symbolic view of knowledge led Cassirer to further analyses of the formal principles that guided the
process of theoretization in the sciences and in the cultural disciplines. The dual principles of ideality and universality, originally uncovered in the context of the exact sciences, were now
evidenced in the development of spoken language, made more

definitive in written language and prescientific, discursive forms of thought.

It now seemed apparent to Cassirer that his efforts to find the philosophical significance and meaning of knowledge should be directed toward an analysis of thought itself. How else could one understand the varying manifestations—scientific, linguistic, mythological, and artistic—of this characteristic logical trend of symbolic envisagement? Thus, out of the unitary matrix of perceptual experience man seemed able to construct a rich and diverse set of representational patterns of meaning.

Cassirer's two greatest concerns now were to pursue the philosophical significance of science and to preserve for language, myth, religion, and art a legitimate epistemological function. The further his logical principles penetrated the various cultural and symbolic domains, the clearer it became that each of those diverse strands of knowledge must be traced back to its source. It could be said that as a result of this phenomenological analysis of knowledge there seemed to exist two general logical principles which undergird the character of symbolic expression. But this was not enough. To leave the issue, satisfied with a mere reporting of an existent state of affairs, one would probably open the door to new metaphysical assertions. For example, it could be asked whether "ideality" and "universality" were principles independent of human control; were they in nature, outside, and imposed upon man by the character of things? Perhaps one could hypothesize that these principles exemplified the coercive power of thought, imputing to a materialistic and heterogeneous world a quality of things that is reflective only of the nature of man's mind.

In the third and final volume of *The Philosophy of Symbolic Forms*, Cassirer bypassed a direct examination of these philosophical questions by raising the investigation to a new empirical level. This volume, completed in 1929, developed into a phenomenology of knowledge in which the concerns of all his writings on knowledge were synthesized and joined to a new trend in Cassirer's thinking. It marks the beginning of the third and final phase of Cassirer's pursuit of the philosophical meaning of exact knowledge: a shift from a concern with theory (science) and thought (language, myth) to the perspective of man himself. True, this shift is nascent to an extent implicit, yet the germ is unmistakable and the direction clear.

Cassirer's encounter with the writings of Henry Head provided the most significant impetus toward this turn. Head's work with the problem of aphasia had been appearing in English publications since 1920 and was climaxed in the publication of his *Aphasia and Kindred Disorders of Speech*.[1] Cassirer thus became aware of the significance of aphasia subsequent to the publication of the first two volumes of *The Philosophy of Symbolic Forms* (1923 and 1925). He found in Head's research a new and exciting substantiation of his own anti-empiricistic and -substantivistic views on the nature of symbolic thought.[2]

This new perspective on thought now concentrated on pathological defects in the use of language that revealed deficiencies in not only one human function but in the "total behavior and mental state" of the person. Cassirer confronted each new ramification of his theoretical problem with an empirical, scientific, and factual study of the problem in order to establish more firmly his overriding concern—a theoretical understanding of the issues. This is notable especially after he became involved with the work of Adhémar Gelb and Kurt Goldstein on aphasia in the Frankfurt Neurological Institute. Goldstein, who had a typically continental respect and sympathy for philosophical issues, invited Cassirer to observe and take part in the experimental activities at the Institute.[3] It was here that Cassirer obtained a vivid awareness that the abstract and philosophical concern for the nature of scientific theory ultimately had to be resolved by an examination of the nature of human thought and language.

1

Like every other scientific issue, aphasia too, which was not unknown in earlier eras, had had its varied interpretations. During most of the nineteenth century the problem had been viewed through the writings of sensationalistic, associationist psychologists.

> Here the psychological concept of the impression as defined by Berkeley and Hume was in all seriousness elaborated anatomically and physiologically. Every cell or group of cells in the brain was held to be endowed with a special ability acquired by experience, to receive and preserve certain impressions and then to compare these stored-up visual, auditory, and tactile images with new sensory contents. The old metaphor of the *tabula rasa* reappears: Henschen, for example, in explaining how we learn how to read, declares that certain letters or engrams are stamped on our brain cells, very much "as the form of the seal ring is imprinted on the wax."[4]

The noted Hughlings Jackson, one of the pioneers in the study of aphasia, extended considerations of the pathology of speech to a more general frame of reference, connecting it with disturbances of optical and tactical recognition.[5] Thus it was seen that aphasia disturbances did not affect "speech as an isolated act, but rather that every change in a patient's language world always brings about a characteristic change in his behavior as a whole—in his perception as well as in his practical, active attitude toward reality."[6] Jackson's investigations of aphasic disorders in language led him to the insight that the loss of capacity to use words does not mean merely the loss of a certain vocabulary.

This unreliable type of verbal inventory caused him to examine the function of words as they were used. The results of these studies seemed to divide speech into two distinct categories: emotional utterances and statements or expositions. In aphasic disorders the latter were affected in far greater degree than the former. "Superior" speech could then be considered to have propositional value, whereas "inferior" speech disclosed only inner emotional states and feelings. According to Jackson, all the truly intellectual powers of language, all that is accomplished for thought, were contained in that power of statement or predication.[7]

Head, the developer of the most systematic treatment of the problem of aphasia, quoted Jackson as stating the following view of language: "It is from the use of a word that we judge of its propositional value. The words 'yes' and 'no' are propositions, but only when used for assent or dissent; they are used by healthy people interjectionally as well as propositionally. A speechless patient may retain 'no' and yet have only the interjectional or emotional, not the propositional use of it; he utters it in various tones as signs of feeling only."[8]

In developing Jackson's conception, Head expanded the notion of the propositional use of words to the faculty of "symbolic expression and symbolic formulation."[9] The first use of the term "symbol" in regard to aphasia was made by Wernicke in 1874 as "asymbolia."[10] Head believed such symbolic disturbances were to be found in many other human achievements and activities. According to him, most of our voluntary movements and activities in reality contain within themselves a symbolic element which in his opinion needed more classification and recognition for their unique character to be understood.[11] His major work, published in 1926, *Aphasia and Kindred Disorders of Speech,* marked the summation of a lifetime of empirical and theoretical research into the function and significance of these pathological cases and their meaning for our understanding of human behavior. His conclusions on the significance of aphasia as it affects symbolic behavior were as follows:

> By symbolic formulation and expression I understand a mode of behavior, in which some verbal or other symbol plays a part between the initiation and execution of the act. This comprises many procedures, not usually included under the heading of the use of language, and the functions to be placed within their category must be determined empirically; no definition can be framed to cover all forms of action which may be disturbed at one time or another according to the nature and severity of the case. . . . But

any act of mental expression, which demands symbolic formulation, tends to be defective and the higher its propositional value the greater difficulty will it present. . . . Any modification of the task, which lessens the necessity for symbolic representation, will render its performance easier.[12]

At the time Head was thus synthesizing his researches, Adhémar Gelb and Kurt Goldstein were similarly turning their attention to the aphasia problem. Goldstein's work in the Frankfurt Neurological Institute, as we have mentioned, was readily available to Cassirer, and it was here that Cassirer acquainted himself with a rich store of concrete examples of the disruption of symbolic behavior which is manifested in aphasia. Thirty-one years after Head had summed up his thinking on the problem, Goldstein, in 1957, at the apex of a significant career, could still make the same general confirmation of Head's earlier position.

As I could demonstrate by a great number of examples, a defect due to a brain cortex lesion cannot be understood or even correctly described as long as one looks at it as an isolated phenomenon. It must be considered from the viewpoint that all human behavior, normal as well as pathological, is governed by a basic motif of organismic life, by the trend of the organism to actualize itself, its capacities, its nature as well as possible under the given circumstances. This self-actualization includes the coming to terms of the human organism with the outer world, especially with its fellowmen. Language appears in this aspect as a special means to effectuate this coming to terms in the process of self-actualization. The change of the personality which accompanies some defect of language reveals the close relationship between the self-realization of man and the use of language, and a study of this relationship is thus suited to affording an insight into the nature of human language itself.[13]

As we pass from Jackson to Head, Gelb, and Goldstein, we note their persistent observation that, in aphasia, the pathological consequences of behavior and language in practice disclose two states of thinking. One level is practical; it involves the everyday concrete use of words. This stratum does not reveal the need for symbolic thinking in the use of language. On the other level we find the full and rich use of representational and propositional forms of speech which constitute the regular mode of expression. In aphasia the latter level seems absent.[14] This does not only occur in the restricted sense of aphasic disturbances of speech (amnesic aphasia). It is also noted in those disorders called *agnostic*, which affect tactile and optical perception, and also in *apraxia*, which denotes pathologies of action or of various activities. Cassirer cites Heilbronner as noting that these disorders all represent a variety of a general form or picture of the disease and that "the singling out of aphasia as an independent pathological grouping may be explained and justified more by practical requirements than by any purely theoretical considerations."[15]

Throughout his writings, Goldstein emphasizes one of the most important empirical confirmations of this theory—that a language loss cannot merely be associated with a physiological impairment of perception or with the disappearance of phonetic images. In the amnesic aphasiac the loss of the capacity to designate universal color names is the most obvious exemplification. One such patient could never spontaneously use names such as "yellow," "blue," "green," or "red." Strips of colored materials meant nothing to him when he was asked to sort them out by color. Although he was as able as a normal person to differentiate color differences, he was not able to name the color and he could not classify colors according to any principle of color families. The patient could choose colors on the basis of color tone or brightness, and could always pick out colors when they were linked with certain concrete objects, such as "ripe strawberry," "sky," "chalk." Also, through certain memorized trains of association, the word "red" could be translated to "blood" and vice versa. However the generic word "red" still remained meaningless as compared to its universal significance for a normal person.

Gelb and Goldstein summed up their conclusions concerning this problem after examining a large number of such cases.

In sorting the colors the normal individual is impelled in a certain direction by the instructions given him. In accordance with the instructions he considers only the basic color of the model, regardless of its intensity or purity. The concrete color is not taken in its purely singular facticity, but more as a representation for a certain color category, redness, yellowness, blueness, etc. Let us designate this "conceptual" . . . attitude as a "categorical" attitude. The [aphasiac] patient is more or less lacking any principle of classification because for him this categorical attitude is impossible or impeded.[16]

The examples of color name amnesia as well as the other amnesic examples given by both Head and Goldstein indicate that, where the functional loss in language and perception affects the entire behavior of the individual, it actually causes the individual to deal with experience on a new level. The patient is now tied to the concrete, the discrete, and the completely sensuous. He lives within the confines of momentary impressions. Impressions which heretofore directed the mind to greater integrated totalities of meanings now are exhausted in the receptive moment.

Language and perception thus are both grounded in an original functional unity. This goes far beyond the view that experience is compounded of the discrete sensuous moments that are purely perceived by our sensory organs. On the contrary, perception immediately organizes its materials into relationships which common-sense language in a sense formalizes and secures. The child's concrete use of language, as noted by Stern and Bühler, is cited by Goldstein and Scheerer to demonstrate that this verbal specificity indicates an early ontogenetic level of integration and is gradually superseded by the more categorical or abstractive organization of experience, much as the expressive moment of perceptive experience is surpassed and shaped into certain definite and formal modes—art, myth, and religion.[17] "The categorical attitude and the possessions of language in its significatory function are expressions of one and the same basic attitude. Presumably neither of the two is cause or effect. It seems to us that the disturbance which gives rise to all the symptoms we find in our patients consists in an impairment of this fundamental attitude and a corresponding lapse into a more primitive attitude."[18]

The reversion to a more primitive intellectual level does have an analogous anthropological and linguistic setting, which Cassirer duly notes. This fixation of the pathological upon the concrete and sensuous brings to mind the Ewe language, in which the word "unripe lemon" designates the color green and "ripe lemon" the color yellow.[19] In general, starting with concrete designations, language gradually opens a path toward purely relational and abstract significatory expression. Primitive linguistic formation usually differs from the higher stages of language formation in the former's diversity, in the extraordinarily rich particularity of the concepts, in its ever shifting unities of meaning, in its lack of unitary conceptual crystallization. Any process—sitting, walking, eating, or drinking—on this level of language usage, is designated by a special word used according to certain very specific rules, depending on the exact circumstances of the process.[20]

Speech pathology displays much the same rooting in the sensuous particular.

It has been generally observed that aphasiacs who no longer know how to use certain words and sentences with a purely objective representative intent can employ them correctly as soon as they take on a different meaning within speech as a whole, as soon as they become an expression of affectivity and emotion. And at the same time a kind of shift occurs in the sphere of representation as well: abstract terms are replaced by concrete ones; universal terms give way to particular and individual ones; and thereby speech as a whole, compared to that of the normal individual, takes on a predominantly sensuous coloration. Linguistic concepts expressing a purely intellectual relation and determination give way to others which bear a kind of "sensitive" stamp; picturesque terms become prevalent; while all purely significatory expression is more or less inhibited.[21]

2

At this point Cassirer realized that his investigation into the nature of aphasia had some significance for a renewed evaluation of the traditional position of sensationalist psychology and epistemology. He felt that given the scientific evidence the relation between language and perception had been established. For the configurations and intuitions of the perceptual world to have meaning, they must first be immediately ordered into a unitary intellectual vision. This perceptual Gestalt, on the other hand, acquires stability and permanence only when fixed in the linguistic sound. "What was intended and begun by sense perception is concluded by linguistic meaning."[22] The historic assertion that language is a conventional artifice arbitrarily created by thought to bedevil reason, as had been argued by sceptics of linguistic realism from Carneades and Occam to Hume, and on through the modern empiricists, was claimed by Cassirer to have been an error.

The universalizing concept which language builds into its relational structure is not a purely logical conventionality, nor is it an assertion of the ontological existence of such a universal "thing." It is rather a manifestation, a product of man's inherent organic ability to see meanings in his perceptual experience, meanings which relate various strata of this experience. Perception itself already encompasses *Gestalten* rather than associated processes producing ideas:

> . . . . the dividing line that is here drawn between the worlds of perception and language should actually be drawn between the worlds of sensation and perception. Every conscious, articulated perception presupposes the great spiritual crisis which, according to the skeptics, begins with language. Perception is no longer purely passive, but active, no longer receptive, but selective; it is not isolated or isolating, but oriented toward a universal. Thus perception as such signifies, intends, and "says" something—and language merely takes up this first significatory function to carry it in all directions, toward realization and completion. The word of language makes explicit the representative values and meanings

that are embedded in perception itself. And, on the other hand, the thoroughly individual, singular perception which sensationalism and with it the skeptical critique of language sets up as a supreme norm, an ideal of knowledge, is essentially nothing more than a pathological phenomenon—a phenomenon which occurs when perception begins to lose its anchor in language, and when its most important access to the realm of the intellectual is thereby closed.[23]

Cassirer also examined two other kinds of pathological impairments—agnosia and apraxia. Both tactile and optical agnosia are manifested when patients are unable to generalize on the basis of their sensory experience. A patient suffering from tactile agnosia, for example, when given a coin is able to distinguish its purely sensory qualities, that is, cold, smooth, and heavy, but not that it is a coin. Persons suffering from optical pathologies, so-called "psychic blindness," fall into patterns of disturbance which, for example, might cause them to mistake an umbrella for a leafy plant on one occasion and on the next, for a pencil.[24]

Apraxia, following Liepmann's definition, includes "every disturbance in voluntary movements pursuing a definite purpose, provided this disturbance is conditioned neither by lack of mobility in the pertinent part of the body nor by deficient perception of the objects toward which the action is directed. In apraxia the mobility of the limbs is retained, the impairment is due neither to paralysis nor paresis, and the patient's mistakes in action are not brought about by a failure to recognize objects."[25]

Liepmann distinguishes two varieties of apraxia—motor apraxia and ideational apraxia. In the former, there is no pathological change in the plan of action or in the intention. While the plan of movement has been correctly conceived, it cannot be translated into action. The part of the body affected by the ailment refuses, in a sense, to obey the will. It will not go in the direction prescribed by consciousness.[26] In ideational apraxia, theory and practice or will and action function closely together. We find that the actions are bound inextricably with the form of thought and consciousness. Here, the coordination between the intention of the total action and the ability to organize this general intention into a series of individual separate steps has in some way been affected.[27]

Cassirer cites the case of an individual who could perform the operation of knocking at the door or hammering a nail as long as the objects in the action were in some kinesthetic contact with him. Once contact was withdrawn, the patient, standing a foot away from the objects, was unable to carry out the movement.[28] Again, in other cases, patients could perform certain actions if correlated activities were demanded, such as raising the hand to take an oath, when ordinarily a voluntary raising of the hand could not be achieved. Such blocks to volition were overcome when the person was angry and made a threatening gesture.[29] These actions then are meaningful only to the extent of their relationship with the concrete situation; in a sense, they have become fused with the situation. They cannot be liberated or performed independently. "What impedes this free use seems to be not so much the patient's inability to create a sensuous, optical space as the medium and background for his movement, as lack of any free play for them. For this latter is a product of the 'productive imagination': it demands an ability to interchange present and nonpresent, the real and the possible."[30]

The contrast between normal man, who lives in a world of mediated signs and meanings, and the apractic, who must aim at a direct goal for his actions, is exemplified by the fact that many apractics who under ordinary conditions could not pour themselves a glass of water when asked, do so spontaneously when thirsty. The multitude of individual and mediated symbolic steps between the request to pour the water and the actual act is too overwhelming for them to integrate and respond normally. Some of Head's patients when playing billiards could do well when aiming at the ball they intended to hit but could not conceive of "playing off the cushion" to hit a ball or to use a third ball as an intermediary for the hit.[31] Such mediators are always symbolic: one must tear oneself from the presence of the real object and freely actualize an ideal aim that can be arrived at only in thought. But this symbolically mediated attitude is the same one which conditions the use of language as we know it.

The impairment of this attitude not only inhibits and impedes the use of language but also every other activity—reading, writing, etc.—which deals with signs of things and their meanings rather than with the objects themselves.[32]

As Cassirer put it more generally, it is not the mode of seeing that is impaired, but the "form" of vision which affects the whole of the plan of action. Every kind of movement demands a definitive kind of vision—an intellectual anticipation, a preview into the future and all that is merely possible.[33] The symbolic mode of vision, the intellectual form of man's actions is permeated by what we can call will. Intellection without will cannot be sustained. Thus the Scotist and the Thomist controversy over the primacy of will versus intellect is resolved in the symbolic act. By contrast, the aphasiac is bound to the concrete and habitual. Will transcends this bond. It reaches to the future into the realm of the possible by placing the possible before an individual in a symbolic act. All action (action that is conditioned by thought) occurs in the light of an ideal plan which anticipates the continuity of the action and assures it unity, cohesion, and continuity. The more a plan of action is influenced by this intellectual voluntariness, the richer the dynamic will be, the more ideational the form of activity will be. Thus any action by man becomes significant, not because of the end product, but because of the mode of formation of the plan and the process of the action. This union of idea and will becomes further significant because it brings to man the freedom to prepare and fulfill his varied capabilities for dealing with experience.[34]

The aphasiac's disability in dealing with categories of thought such as the ideal, the possible, the future is given interesting perspective by an example of primitive thought. Cassirer cites a report of Karl von den Steinen on the Bakairi, which on the surface seems to parallel the inability of aphasiacs to repeat a phrase not true to their concrete experience. Von den Steinen noted that his native interpreter found it almost impossible to "translate sentences whose content for any reason seemed senseless or impossible to him: he would shake his head and decline to render such sentences."[35] This inability, of course, is not due to any physiological incapacity but it does emphasize the important role of social convention on both thought and language. In this case, it shows how the will can be paralyzed and can succumb to the common-sense social usage of language.

While we can accept, on this level of human existence, Whorf's view that language is the shaper of man's vision of the world, it is only because primitive man so wills it. In this particular cultural context man has allowed himself to become a passive observer of and conformer to his environment. This is reflected in his language, whose intuitive morphological structure now becomes rigidified. Given a more fluid social environment and the consequent attitudes of mind thereby engendered, the Bakairi too would find the linguistic means of intellectually transcending the given cultural norms of thought.

The aphasiac's inability to use linguistic analogies and metaphors correctly, on the other hand, can be traced back to his psychic difficulties, for the use of metaphors involves the use of words in new shades of meaning. This involves the constant shifting of position with regard to these varied meanings, an impossible task for the aphasiac. He is forced to cling to the present, the sensuous, and the concrete. His language is likewise conditioned by his intellectual attitude: " . . . without this support he is rudderless—he cannot venture out on the high seas of thought, which is a thought not only of realities but also of possibilities."[36] Thus he can only express what is actual and present, never the imagined or the possible. To think and speak of the future, one must break with the present, one must disregard it and look toward an ideal goal.[37]

All of these cases, in the variety of shadings of the disorders, together with the diverse symptomology, are unusually extensive in range. Cassirer agreed with Head in that, within this range of pathological disturbances, no impairment of any particular faculty can be shown to have occurred; " . . . it is not the loss of a faculty that we have here, but the transformation of a highly complex psychic and intellectual process. According as the change affects this or that characteristic phase of the total process, very different pathological pictures may arise, none of which need resemble the next in its concrete traits and symptoms but all of which are nevertheless linked together insofar as the change or deviation in all of them points in the same direction."[38]

3

Cassirer, at this stage in the development of his thought, was still wary of attributing a substantialistic or material substrate to the process of symbolic thought. For the purpose of Cassirer's philosophic quest, the question as to the substantial reality of these disfunctions in symbolic capacity is unimportant. The importance of this empirical evidence lies rather in its meaning for human culture: "we must strive to bring the teachings of pathology, which cannot be ignored, into the more universal context of the philosophy of culture."[39] Cassirer is still very much the Marburg neo-Kantian concerned with function and form. The other road which leads to epistemological realism in scientific thought contains all the pitfalls that have brought philosophical disaster to two thousand years of metaphysical speculation.

This philosophical restraint did not inhibit Cassirer's authority on aphasia, Kurt Goldstein. Goldstein located the area of disturbances of the symbolic functioning of man in the workings of the cortex, especially in the frontal lobes of the human brain.[40] He characterizes the changes in function which an aphasiac undergoes as a regression to a more primitive level of behavior. This can be called the concrete attitude, as contrasted with the abstract attitude.[41]

In the concrete attitude, things, words, and actions relate as objects of specific practical activities. The apractic uses the spoon or cup properly during the meal but otherwise their use becomes a meaningless series of sense perceptions. He displays a system of behavior which is closer to life. In concrete performances a reaction is determined directly by a stimulus; the quasi-authoritative responses that are awakened are all that the individual really perceives. The individual's attitude is thus extremely passive, as if it were not he who had the initiative.

In "abstract" performances, an action is not determined directly and immediately by a stimulus or configuration, but by the account of the situation which the individual gives to himself. The performance is thus more a primary action than a mere reaction. And it is a totally different way of coming to terms with the outside world. The individual has to consider the situation from various aspects, pick the aspect which is essential, and act in a way appropriate to the whole situation.[42]

Cassirer likewise stressed the passive, almost instinctual and animal-like character of the aphasiac's pathological behavior. An animal is certainly unable to thrust his environment back from itself and take account of the situation, to contemplate it at a distance. It can never create this distance or represent it to itself through signs. It lives its environment.[43] Its only assertions lie in accurate responses to stimuli which strike it from the outside. Each response comprises a predetermined temporal sequence of actions. Instinct in general merely signifies that in certain situations, under fairly similar conditions, the animal will respond to an external series of stimuli in exactly the same manner. The animal cannot be aware of its own actions nor can it represent them in consciousness. The unraveling of one phase or stage of action from another necessitates no subjective apprehension of the process by the animal ego. Truly, the animal is a captive within this sequence of action. It cannot voluntarily say "no," nor depart from its prescribed course by conceptualizing any of its activities. And finally it has a radically limited ability to anticipate the future, to imagine or prepare for an eventuality that has not occurred.

Only man forms new modes of action out of new temporal visions. Man distinguishes, chooses, and judges, always creating in effect an extension of his own personality into the future. "What was previously a rigid chain of reactions now shapes itself into a flowing and mobile, yet self-centered and self-contained, sequence in which every link is determined by reference to the whole. In this power of 'looking before and after' lies the essential property and basic function of human 'reason.' 'Sure he that made us with such large discourse, / Looking before and after, gave us not / That capability and godlike reason / To fust in us unus'd.' "[44] Man acts both discursively and intuitively. He clearly differentiates his goals beforehand, then reunites the elements in a unitary vision. It is this temporal differentiation and integration which gives action its symbolic imprint. It demands free movement, yet it implies a directed movement toward a unified goal.[45]

Man, in freeing himself from the compulsions of sensory drive and immediate need, is driven by new forces which he seeks to control and conquer. But now they are theoretical or conceptual rather than practical or biological. The description of the aphasiac as one drifting toward an instinctual animal-like level of behavior is not quite true. The regression in which he is involved is only partial, for he cannot fall back on any prescribed instinctual behavior. Man is bereft, by and large, of those clearly defined instinctual patterns of reactivity to external signals. Basically the aphasiac is limited to only the most automatic of speech responses and those rudimentary behavioral and perceptual abilities learned at an early age. All that is left are patterns that have been retained by the brain on a simple level of functioning, somewhat like the more autonomous operations of the body.

The aphasiac or apractic seems to have been thrust one step backward along this path which mankind had to open up by slow steady endeavor. Everything that is purely mediated has in some way become unintelligible to him; everything that is not tangible, not directly present, evades both his thinking and his will. Even though he can still apprehend and in general correctly handle what is "real," concretely present, and momentarily necessary, he lacks the spiritual view into the distance, the vision of what is not before his eyes, of the merely possible. Pathological behavior has in a sense lost the power of the intellectual impulse which forever drives the human spirit beyond the sphere of what is immediately perceived and desired. But precisely in this step backward it throws a new light on the general movement of the spirit and the inner law of its structure. The process of spiritualization, the process of the world's "symbolization," discloses its value and meaning where it no longer operates free and unhindered, but must struggle and make its way against obstacles. In this sense the pathology of speech and action gives us a standard by which to measure the distance separating the organic world and the world of human culture, the sphere of life and the sphere of the objective spirit.[46]

What Cassirer is telling us is that the cultural world represents an irrevocable break with its organic tradition.[47] It is man's creation, a biologically unique intellectual and ideational vision of experience. Science, art, myth, and common-sense language are pictures of experience, symbolically created by man to give meaning to his life. If we wish, therefore, to understand all of life, the material as well as the ideational, we must view human culture not as a response to an organic need or superficial involvement with matter, but as part of the symbolic mediated and conventional world of cultural meanings.

But these meanings are not random or haphazard creations. Though freed from the strict instinctive determinism of the animal, man must still confront his own nature. Although the objectivity of self-analysis is difficult to establish, the trends in man's thought as exemplified both in history and in contemporary culture do represent a process having meaning. The problem is to ascertain what the meaning might be.

We should note first that Cassirer divested knowledge both of its realistic and reductive metaphysical attachments. He showed that the form of scientific thought was a matter of convention and could be qualified only by its predictive and logical functional characteristics. The logical element he noted to be rooted in a perennial demand of thought on the form of experience. Further he found that the trends toward ideality and universality were logical characteristics of other realms of thought, as, for example, in the development of language systems. They were characteristic also of normal perceptual and linguistic behavior. Finally, through his analysis of aphasiac pathologies he came to the conclusion that this logical trend of human thought took on a new and more general significance because it set itself off so sharply from nonhuman organic behavior.

The question that yet needs answering was how Cassirer intended to establish a set of non-ontologically rooted *a priori* principles that would constitute the logical ground for all discursive forms of symbolic thought, from ordinary perceptual behavior to theoretical physics. Could this be done without succumbing to the substantialism inherent in Kurt Goldstein's attribution of a specific source for abstract thought in the cerebrum and frontal lobes of the brain? Is Goldstein here repeating that original Kantian error of accepting Sommering's realistic attribution of the analytic and synthetic activities of mind to the chemical structure of the brain?[48]

7

Man as Symbolic Animal:

The Theoretical Issue

CASSIRER'S IDEAS ON THE LOGICAL SIGNIFICANCE OF man's seemingly universal cognitive and abstractive strivings did not coalesce for some fifteen years after the publication of the three-volume *Philosophy of Symbolic Forms*. During these years Cassirer embarked on a long and hectic hegira which took him from Hamburg to Oxford and Göteborg, thence to Yale and finally Columbia. Considering the events of the time he was amazingly prolific. When Cassirer did return to the problem of cultural knowledge, in his *An Essay on Man* (written in English), he gave the question a new turn. In this volume, rather than pursuing the abstractive demands of human thought through psychological or morphological analyses in materialistic and substantialistic terms, he turned to the historical and evolutionary dimension.

While the argument of *An Essay on Man* is to an extent a restatement of his more detailed *The Philosophy of Symbolic Forms*, it is no mere summary.[1] A wholly new perspective issues out of this work. It is an extension completely congruent with Cassirer's mode of research and is documented in much the same thorough way as were those epistemological areas he had previously analyzed.

1

The new approach was predicated on the inescapable conclusion that symbolic thought denoted a generic characteristic of human behavior, a characteristic which had its roots in a conception of human nature. As developed from the writings of Uexküll, among others, the symbolic quality of thought, shown to be partially denied to aphasiacs, was demonstrated to be completely nonexistent in the biological world.[2] Symbolism, as expressed in philosophy, science, language, and even perceptual thought, is a manifestation found in man alone. The behavior of animals, no matter how elaborate, is instinctual. Their signal behavior, mediated through specific perceptual stimuli, has clear-cut adaptive goals: reproductive efficiency, natural selection by the environment, and eventual stasis with the environment. The switch from a system of signal responses to the symbolic level of thought and action represents a qualitative shift from the gross, if secure, substantialism of materially oriented behavior, to a new, perhaps precarious and even inexplicable, level of organic existence.

> Symbols—in the proper sense of this term—cannot be reduced to mere signals. Signals and symbols belong to two different universes of discourse: a signal is a part of the physical world of being, a symbol is a part of the human world of meaning. Signals are "operators"; symbols are "designators." Signals, even when understood and used as such, have nevertheless a sort of physical or substantial being; symbols have only a functional value.[3]

"In short, we may say that the animal possesses a practical imagination and intelligence whereas man alone has developed a new form: a *symbolic imagination and intelligence.*"[4]

Symbolic intelligence then cannot be conceived to function on a substantial and biological level of thought, as a product of material, survivalistic needs. Just as Cassirer had heretofore turned aside explanations of knowledge in terms of entities—noumena, the raw given of sensations, material substance—so too he now rejected all arguments for explaining man in terms of anterior existing standards of objectivity against which human behavior must be measured. By contrast, symbolic thought must be defined in terms of its own functional or behavioral characteristics. And yet we cannot avoid the problem of historical explanations. Symbolic thought is a historical product of nature and must be so understood and explained.

> Man has, as it were, discovered a new method of adapting himself to his environment. Between the receptor system and the effector system, which are to be found in all animal species, we find in man a third link which we may describe as the *symbolic system*. This new acquisition transforms the whole of human life. As compared with the other animals man lives not merely in a broader reality; he lives, so to speak, in a new *dimension* of reality. There is an unmistakable difference between organic reactions and human responses. In the first case a direct and immediate answer is given to an outward stimulus; in the second case the answer is delayed. It is interrupted and retarded by a slow and complicated process of thought. At first sight such a delay may appear to be a very questionable gain. Many philosophers have warned man against this pretended progress. "L'homme qui médite," says Rousseau, "est un animal dépravé": it is not an improvement but a deterioration of human nature to exceed the boundaries of organic life.[5]

Cassirer made no attempt to introduce principles not in consonance with the empirical evidence. Evolution did not grind to a halt with the introduction of man to this planet. Yet human life did introduce a wholly novel factor into the history of organic forms. Human perceptions and actions are not the results of clear biological needs and satisfactions. With man were introduced propositions and abstractions, imagination and illusions.[6] Thus, Cassirer asserted, man is the product of a new mutation in life, one which departs radically from the traditional direction of organic evolution, but not from the form or structure of the process itself. Nature has always allowed for innovation, for new creatures which transcend the existing forms of life.[7] In some similar way, man too must have been picked up by the forces of nature and deposited onto this new and perilous road which it is his destiny to traverse.

The sketch of human nature which Cassirer drew in his *An Essay on Man* was entirely consistent with his position on knowledge. His position as it derived from the nature of philosophic and scientific thought, from language and common sense, led to this logically *a priori* invariant—human nature. Here the search for meaning and context in knowledge could go no further. Knowledge reflected no anterior existing reality. Nor was it reducible to some ultimate and discrete desideratum of facticity. The structure and direction of knowledge could be clearly traced to the demands that thought makes of experience. The variety as well as universality to be found in the forms of knowledge is likewise traceable to a freedom that thought has been able to apply to experience. This freedom is seen as being rooted in the unique character of man himself, which is in turn traceable to the great alteration in organic behavior now manifested by man alone.

From the standpoint of the problem of knowledge Cassirer's new conception, human nature, provides a unique architectonics. Not only is Cassirer saved from pointing out any one solution as to the source of knowledge, whether in a subject matter such as mathematics, physics, or in the ontological absolutes of perceptions, ideas, or pure form, but also he is rescued from the fragmentary relativism in knowledge suggested by his diverse symbolic forms. Focusing on the problem of human nature can assist us in turning back to the disciplines of knowledge to understand both the logically universal and specific dimensions in each.

Though the consistency of his neo-Kantian philosophical position was borne out by the facts in each succeeding discipline to which Cassirer addressed himself, his new *a priori* invariant presented a special intellectual problem. The uniqueness of human behavior, thought, and culture, examined from a gross biological perspective, is of course evident. In an examination of even the most primitive contemporary peoples or their ancient paleolithic counterparts, one can perceive the special qualities of the human condition. But to perceive and note this historical alteration is not to explain it, especially when one is faced with the incontrovertible evidence given in the structure of evolutionary theory. The logical demands which Cassirer himself established necessitate that no empirical domain be split in two by incommensurable theories. Can man be natural as well as cultural if we postulate in nature a new beginning, a grand mutation, in which the laws governing behavior are written in an entirely new orthography?[8] Further, how strong can the neo-Kantian and instrumentalist approach to knowledge be if founded on invariant principles whose truth is itself subject to question?

It is unfortunate that Cassirer was not able to pursue these questions to their resolution. That he was aware of the logical predicament is shown in the following passage:

> That symbolic thought and symbolic behavior are among the most characteristic features of human life, and that the whole progress of human culture is based on these conditions, is undeniable. But are we entitled to consider them as the special endowment of man to the exclusion of all other organic beings? Is not symbolism a principle which we may trace back to a much deeper source, and which has a much broader range of applicability? If we answer this question in the negative we must, as it seems, confess our ignorance concerning many fundamental questions which have perennially occupied the center of attention in the philosophy of human culture. The question of the *origin* of language, of art, of religion becomes unanswerable, and we are left with human culture as a given fact which remains in a sense isolated and, therefore, unintelligible.[9]

2

Having taken us from the abstractive heights of epistemological theory in philosophy, then through the various theoretical models of physical science, culture, and perceptual thought, finally arriving at a base from which the great spectrum of human thought could be examined in the context of a theory of human nature, Cassirer seems to have concluded his logical search for an *a priori* invariant of knowledge with a paradox. Throughout his perusal of the Kantian theme of the autonomy of thought in the creation of theory, Cassirer had shown that the various reductive metaphysical schemes of sensationalism, mechanism, phenomenalism, and substantialism did not fit into the evidence presented either by the structure or the function of the various scientific disciplines. This was confirmed by the empirical evidence adduced by students of the various cultural disciplines. Thus, in coming to the problem of man and enunciating his general theory of symbolic thought, Cassirer had presumably reached a unique and original point of departure. Once more he would turn to the structure of knowledge in all the multivarious domains of human action.

But in postulating the independence of man from the crude material and instinctual drives of animals, he had broken one of the cardinal principles of the scientific and philosophical world view which he himself had espoused. He had violated the canon of the universality of theory as it applies to all experiential domains subject to its basic premises. In doing so, he questioned the logical ground which makes theory in science at all possible—the continuity of nature. Can man be conceived as a unique creation of nature, a creation in which the laws governing the evolution of animal life are bypassed? No, unless one is willing to accept traditional theistic and ontological explanations of human nature, which would hardly be consistent with Cassirer's naturalism.

The fact of biological evolution is of course not an insignificant rock for any theory to founder upon, even a theory as well buttressed empirically as Cassirer's. Helmut Kuhn, a philosophical ally of Cassirer's, in reviewing *An Essay on Man*, considered Cassirer's neo-Kantianism the euthanasia of the Marburg school.[10] The specific problem Kuhn raised in this regard was Cassirer's underlining of a seeming discontinuity between nature and thought. This was also noted by a number of other sympathetic writers.[11] Perhaps the tersest statement of this concern was made by I. K. Stephens, who chided Cassirer for making man's "will to logic" fundamental—overlooking the "will to live" and neglecting the whole operation of evolutionary adaptation.[12]

Cassirer's logical predicament was not merely the product of an *experimentum crucis* (which turns out in retrospect to have been not so crucial in the first place). It lay rather in the confrontation or apposition of one great theoretical model with another; the adoption of either would inevitably alter the character of a vast number of concepts in a variety of disciplines. As has been noted with regard to other important turning points in intellectual history—the transition from Ptolemaic to Copernican cosmology, from Newtonian electrodynamics to the field theories of Faraday and Maxwell, or from the gravitational theories of Newton to Einstein—one cannot judge the validity of a theory solely by its ability to manage and account for the data of experience. Practically every theory can be manipulated so as to manifest a logical structure which will order the phenomena. It is when two systems of ideas, such as Cassirer's symbolic theory and the traditional views on Darwinian evolution, come to seemingly contrary deductions concerning human behavior, its causes and directions, that one is faced with the possibility of having to abandon one or the other.

It is necessary then to examine briefly the model of evolutionary biology. The implications of this theory for the nature of human knowledge far outspan the immediate issue as to whether human symbolism, in its motivations and significance for thought, is so divergent from animal behavior. The determination of this issue potentially penetrates every intellectual discipline and has implications for even the most prosaic social concerns. A theory on this level of generality is like a beacon which can throw into relief entirely new facts and meanings heretofore quite differently accounted for.

There were of course evolutionary models developed by such men as Lyell and Spencer both before and contemporaneously with Darwin. But with the postulation, in *Origin of Species* in 1859, of the specific logical mechanism through which the purported process presumably took place, all alternate models were disregarded. Darwin's trinity of mutational changes, adaptation, and natural selection, acting mechanically and ineluctably, was powerful enough to persuade investigators in a variety of fields to develop their facts so as to universalize this principle and apply it to the study of man and society. Witness the bitter fruit born of the early "social Darwinist" views of William Graham Sumner. Here the concept of natural selection as enunciated in the slogan of the survival of the fittest was applied so as to give intellectual sanction to the social status quo of privilege and power.[13]

In the new discipline of anthropology, Edwin B. Tylor and Lewis Henry Morgan seized upon the obvious empirical fact of social evolution to signify that, as in biological evolution, there existed laws of social progress that could explain the evolution of culture. The development of civilization from primitive to modern times was seen as a unilinear process having an inevitable and irreversibly progressive character.[14] The great and obvious differences between the human world of social and cultural existence and the organic world of drives and instincts were not seen as insurmountable historical problems. Rather they presented a challenge for theory to link the two realms and thus satisfy a law of logical order. In this respect Darwin's theory acted in the same epistemological manner as did Newton's, Einstein's, or modern quantum theory. It directed the mind to ever wider sets of empirical data to allow scientists to see that what was thought to be part of a unified set of data was essentially diverse and heterogeneous material, or the converse. In the case of the political and economic thought of Marx—a well-developed and well-articulated theory in another realm—support was lent to Darwin's theory by dint of the materialistic analogy. It gave Marxism an additional measure of credence that it would not have otherwise gained at that time.

The shift from the empirical and analogical phase of Darwinism to a deductive and experimental form more in consonance with the models employed by physicists took place about the turn of the twentieth century with the contributions of Gregor Mendel, Ivan Pavlov, and Thomas Hunt Morgan.

Darwin had spoken of minute and continuously occurring variations—or sports—in all animal species as constituting the material basis of evolutionary change.[15] Experimental evidence of the character, source, and frequency of these variations was lacking until Mendel, DeVries, and Morgan made their contributions.[16] It was then that these alterations were pinpointed in the genoplasm of each species. The instrumentality of the gene theory of inheritance was soon buttressed experimentally and theoretically. The amenability of this theory to quantitative and predictive techniques of analysis effectively swept away the claims of Lamarck and Buffon as to the importance of the external environment in effecting evolutionary change.

The genoplasm, the source of all evolutionary processes, was thus insulated from outside influences. Changes in the genoplasm were completely random. The role of the environment was to effect a mechanical sorting of those adaptive behaviors over which the creature had no control. Externally the evolutionary process seemed to be objective and deterministic, subject to mathematical and deductive analysis. The nemeses of science—free will, choice, teleology, and vitalism—were effectively removed.

At the turn of the century these convictions were deepened through Ivan Pavlov's experiments with the conditioned reflex. Acting on his knowledge of a creature's basic need for food and its reflex of salivation in immediate perceptual anticipation of the satisfaction of this need, Pavlov was able to elicit the same response with a host of secondary stimuli. Pavlov ultimately demonstrated that a wide variety of behavioral actions and attitudes far removed from the initial stimulus could be traced to and shown to be derived from the basic drive.

> . . . in these conditioned stimuli . . . we have a most perfect mechanism of adaptation . . . for maintaining an equilibrium with the surrounding medium. The body has the capacity to react in a sensitive way to the phenomena of the outer world, even the most insignificant, coinciding even temporarily with the essential become their indicators or, as they may be called, their signalling stimuli.[17]

The ground was now prepared for what was to become one of the most rapid and encompassing theoretical displacements to have been encountered in a discipline.

Darwin's influence in psychology had been evident since the publication of William James's *Principles of Psychology* (1890) and Lloyd Morgan's *Animal Life and Intelligence* (1891). But with the introduction of the term *behaviorism* by John Watson in 1913, we had the official establishment of what could then only be characterized as the scientific school of psychology.[18] The investigation of human behavior, and this included the work of William MacDougall (1908) in social psychology, was dominated by the belief that man's actions could be interpreted, determined, and manipulated through a knowledge of his needs, conceptualized in terms of the adaptive evolutionary model.[19] All members of the school held to a basic demand that their theoretical models not depart from what could be attested to through observation. While these conceptual models were stimulus and response, they had to be thought through in strictly macroscopic and experimental terms. Thus while there existed a consistent demand to see man in terms compatible with the physico-chemical view of matter, and methodologically in mechanical and determinate terms, the cognitive ground which directly affected their experimental and theoretical studies was decisively anchored to the theme of biological survival and those social drives presumed to secure man's earthly tenure.

The increasing complexity of this model eventually led to the sophisticated methodological advances of Clark L. Hull. Hull brought the mechanical and biological schemes together in an attempt through mathematical and deductive means to predict and determine human behavior on the basis of biological needs. His scheme was predicated on the hypothesis that supposedly purposeful actions were truly caused by specific and anterior needs. One only needed to follow carefully the cycle of stimulus-response bonds in their contextual setting and their spatial contiguity to arrive finally at a simple biological drive as initial cause. In a similar manner, by starting with simple activity one could construct a mechanism (robot) which would duplicate the complex actions of men. Indeed, Hull believed that, ultimately, all social behavior could be encompassed within the deterministic mechanical model that was here envisioned.[20]

B. F. Skinner represents perhaps the ultimate resolution of the implications inherent in behaviorism. Skinner's work, predicated again on the assumption that all biological behavior can be traced to specific adaptive drives, focused on an attempt to use the Pavlovian conditioning model on human beings. Starting with pigeons and rats, Skinner attempted to show that human beings could be sheared of behavior that was discomfortingly discontinuous. An orderly universe demanded behavior that was consistent with scientific law. War, hate, turmoil, and exploitation could be avoided by a planned society, Skinner hypothesized in his novel *Walden Two*. Through a process of conditioning which would be applied from birth, craggy and inexplicable human behavior would be programmed away so that peace and happiness would reign in social affairs. In its own way, the theory was not unlike the vision of the social Darwinists, though not illiberal. It was extrapolated whole from the Darwinian assumptions that biological survival depended on the species' ability to satisfy primitive instinctual needs. It was this basic conviction that motivated two generations of psychologists to see in all varieties of seemingly nonbiological and apparently unique human cultural activities merely the refined secondary manifestations of a basic and unilinear tendency for organic maintenance.[21]

The adaptive model was attractive to most psychologists, and for anthropologists and archeologists it became a strong point in their explanation of the character and process of human evolution. The argument was posed that the dominance man has attained in our world is explainable in strictly orthodox Darwinian terms.[22] Man has been successful because he has been able to take advantage of certain basic organic abilities. These opportunities have allowed him to adapt to the environment in such a manner as to insure natural selection. The key to man's survivalistic abilities, while organic, is not physical in the traditional sense.

Man is the ultimate inheritor of an evolutionary trend toward increased brain size which his anthropoid ancestors bequeathed to him tens of millions of years ago. At some time in the past, perhaps barely a million years ago, man's growing brain allowed him to make the first hesitant cultural adaptations—fire and tools—which eventually were to free him from his environmental restraints. The creation of these dual techniques for effecting a more secure livelihood stimulated a dialogue between his evolving genes for brain size, over which man had no control, and his ability to construct more efficient technologies that would fit him for survival. Naturally, not all of our proto-hominids were equally well endowed with the mental ability to handle the new techniques of survival. But those so selected passed on their abilities for cultural modes of adaptation to their progeny. What ultimately occurred was a continuing and reciprocal feedback from good tools to big brains, from bigger brains to improved tools for adapting to the environment. The final consequence was that less well-endowed individuals fell by the wayside.

From the early paleolithic creature with his stone eoliths to the prehistoric neolithic tribes the process of natural selection favored those creatures we have subsequently come to call homo sapiens. The onset of historical cultures marks the end of this challenge and response between man and nature. Man, presumably having met the challenge of the natural world and having assumed a special if universal ecological niche, remains alone, somewhat cut off from the evolutionary process. Henceforth man's evolutionary development would be cultural and historical, not biological and genetic.[23] What must have begun with the random pebble scrapings of a heavy-browed, prognathous anthro-hominid at the onset of the first ice age was completed in the great agricultural evolution at the end of the neolithic, when man finally assured himself of a reliable and continuing food source. In changing from migratory hunting and food gathering to animal husbandry and the cultivation of high-protein grains, man was able to attain in this first economic revolution the higher forms of civilization we have discovered in the archaeological record.

The subsequent historical advance of civilization can be attributed to a continuation of this technological (adaptational) trend toward dominance over the external environment.[24] The above-stated archaeological explanation of human evolution in its earliest cultural phase has been recently supplemented by the so-called super-organicist or culturological position of Leslie White, and the Michigan school of anthropology, which in turn is a modified version of that introduced by the nineteenth-century anthropological evolutionists Tylor and Morgan. While rejecting the ideal of rooting culture in its purported biological sources, the Michigan school strove for an evolutionary theory of cultural change, yet one that could rest autonomously on its own cultural sources. The derivation of these cultural constructs is both materialistic and adaptive. White postulates that the driving force behind cultural change and progress lies in the mobilization, control, and disbursement of energy resources. The advance of culture and civilization can therefore be exactly determined by an analysis of the extent to which each society has harnessed and utilized the energy resources available on the planet.[25]

In general the literature is rich in attempts to explain the evolution and behavior of man in terms of the biological paradigm. Man's political behavior, his wars and aggressions, his purportedly genetically rooted occupational preferences, his abilities and proficiencies in the arts, and finally his theistic and spiritual evolution are traced from various points of view that interpret adaptation and evolutionary theory as part of the ontological nature of things.[26]

The mysterious and unique nature of man's languages also has exerted a fascination for scholars, inviting diverse speculations as to meanings and origins. Indeed, at one time the proliferation of speculations caused the French Academy to reject for consideration all further manuscripts dealing with this subject. The problem of language was crucial to Cassirer in making the transition from the epistemological foundation of science to the problems of culture. He saw language as the most universal manifestation of man's need to envisage experience symbolically. Language was the vehicle through which science became possible. Cassirer's model of language however was idealistic; he could not conceive of it as serving essentially utilitarian purposes.

The Darwinian influence has also extended itself to language theory; there has been no lack of theoretical models employing the adaptational and survivalistic canons of this position. Three general approaches have been articulated, all derived from the same point of view but directing themselves to different aspects of the problem. They might be labeled the genetic, behavioral, and semantic approaches to language.

The genetic view of language runs parallel to the interpretation of man as a tool-making, tool-using creature. It hypothesizes that it was man's ability to fashion more effective weapons and artifacts that allowed him to meet his evolutionary challenge and survive. Likewise with language. Just as the tool became infinitely superior to the fang or the claw, so did language represent a greatly enhanced capacity to communicate as compared with animal cries and gestures. Language allowed men to correlate their actions efficiently for their mutual survivalistic benefit. The meanings that man searched for in language were thus directly applicable and relevant to his problem of physical survival. Writing during the peak influence of behaviorism, Grace Laguna stated the issue as follows:

If we are right, it is to the great superiority of speech over animal cries as a means of *social control* that we must look for the chief cause of its evolutionary origin and development. The primary function of speech is the coördination of the behavior of the individual members of the social group. It is undoubtedly true that in the performance of this function, as speech has become a distinctive mode of response to the objective world, it has likewise brought about an enormous development of individual intelligence. The development of speech has thus acquired an indirect survival value through the increased general capacity of the human individual which it has occasioned. . . . To coördinate the intelligent behavior involved in the use of tools, language is necessary. *Language is correlative to the tool.*[27]

The strength of this view can be seen in anthropological and linguistic theory to this day. On the surface the logic of this position seems irresistible. Man has become the dominant vertebrate animal in our world. Language and technology, and the social organization which is a fundamental derivative of these entities, are the dominant features of the human scene. In that case, both language and tool-using are strongly implicated in man's evolutionary adaptive thrust. The anthropologist Harry Hoijier reflects this contemporary attitude.

> How and in what way man's animal ancestors came to employ language as an aid in cooperative labor we shall never know. In any case, it appears fully certain that language arose as a result of man's learning to work together toward a common end. For whatever reasons, man's primitive ancestors were obliged to acquire such learning, and so they alone of the animals, stumbled upon the tool, language, which more than any other makes cooperative and coordinated activity effective.[28]

Those who espouse behavioral interpretations of language have attempted to eliminate such nonobservable characteristics as thought, insight, ideas, imagination, or other qualities associated with language that would be difficult to reduce to observable and empirically functional activities.[29] The linguist Leonard Bloomfield, for example, tried to show that language acts to stimulate the flow of words and actions in other individuals without implicating the so-called higher symbolic processes of thought. In this interpretation, language could be seen as a trigger releasing quasi-instinctive or automatic behavioral responses: "the speaker has no ideas . . . the noise is sufficient for the speaker's words to act with a trigger-effect upon the nervous system of his speech fellows."[30] This simple schematism runs into difficulty with linguistic behavior in which immediate response is not given, or when responses are given in many divergent pathways which are difficult to analyze from a strict and uniform stimulus-response series. Charles Morris has noted this dilemma.

> In a number of experiments on human beings the results show marked divergencies from similar studies on non-human animals. . . . For non-human beings seldom produce the signs which influence their behavior, while human individuals in their language and post-language symbols characteristically do this and to a surprising degree.[31]

Morris did not resolve the problem by abandoning the model. Rather he modified this rigid interpretation of language functions to include a number of linguistic and thought elements not usually explained through schematisms such as Bloomfield's. "Dreams, fantasies, myths, and the arts function . . . as substitute objects for attaining some measure of satisfaction for frustrated actions."[32] Thus the adaptive image of language is retained by noting that dream, art, and language phenomena that seemingly do not further man's physical survival are in reality the product of nonadaptive behavior, presumably to be eliminated as man moves closer to an adaptive and selective equilibrium in his environment.[33]

The third aspect of the quasi-evolutionary interpretation of language is perhaps the most widely known. It might well be called a philosophy of linguistic therapeutics. Into this category go a host of interpretations and schools, ranging from the semantics of Stuart Chase, C. K. Ogden, I. A. Richards, and Alfred Korzybski to the contemporary philosophical school of linguistic analysis in both Britain and the United States.[34] In this position it is generally assumed that a well-adapted species will exist in a state of equilibrium with its environment and that it will not reflect the chaotic turmoil-filled social panorama that man has created around himself. There is also a concentration on the centrality of language in understanding human behavior and on the belief that the aberrations we therefore see in human thought and action are due to the misuse—the nonadaptive functioning—of language.

The therapies are diverse, extending from suggestions that we clarify our formal language structure, training man to distinguish between language used in an aberrant rather than in a realistic nonmetaphysical and nonmythological mode, to attempts to rectify the vernacular language so as to conform to canons of scientific and logical thought. This has been advocated in a prescriptive sense by Korzybski and Chase. In addition there is the contemporary philosophical practice of concrete and minute analysis of common-sense language to ascertain its various shades of meaning. This is done to enable man to choose the kinds of words and sentences that would serve the social actions that he rationally decides upon.[35] Generally this position recognizes that human language, because of its complexity, is not at all amenable to simple extrapolation from the biological model.

But the difficulties that many scholars encountered in rooting their particular analyses in the theoretical guidelines of evolutionary biology acted as a goad rather than an inhibition to their efforts. For, after all, when one is convinced that there is a logical correlation between animal and human behavior, the mere existence of surface empirical discontinuities must be disregarded in favor of the causal maxim. The purpose of theory is to unite, to see in the heterogeneity of external facts deeper theoretical relationships. And it is this logical imperative that must have priority.

The impact of Darwinism extended itself to every area of study which had to do with living processes. It was not only the adaptational and selectionist paradigm that influenced thought and research. A number of methodological canons carried implicitly within Darwinism constituted the guiding pattern of subsequent work. The entire intellectual apparatus of scientism, objectivity, determinism, quantification, mechanism, and substantialism was incorporated as an integral aspect of every subsequent theoretical postulation. In fact, were one to look beyond Darwin's immediate influence, the omniscient shadow of Newton could be discerned. As yet these methodological aspirations were untouched by Einstein and Bohr.

This impact was clearly seen in the work of Clark Hull, who wanted to spur psychology on to the deductive achievements of the physical sciences. The axiomatic and predictive aspirations of the physical sciences were thence joined to the laboratory studies of the psychologist. Where individual regularities were hard to find, statistical methods dealing with the behavior of large numbers evened out disconcerting rough spots in the model. By dealing statistically with human behavior, it was possible to deal with a large amount of qualitatively neutral information with some predictive regularity. All this seemed to be in complete consonance with the requirements of the scientific method.

There is another, possibly unanswerable, question that we must ask. What kind of social resonances does the mechanical image of man lead to? Does it not postulate the existence of anterior drives that propel man along univocal lines toward the satisfaction of only utilitarian needs? Is it possible that the translation of this model of individual behavior into an intellectual pattern of mass behavior has facilitated the creation of those more execrable social, political, and economic models of recent time?

Is it not possible that when we think always in terms of classes, aggregates, and masses of people, always defined in terms of their gross external needs and predictive behavioral actions, disregarding the problem of values in favor of quantitative and objective scientific categories, we take the inevitable next step? When we do in the social sphere what scientists do in the physical disciplines—to program, predetermine, manipulate—are we not on the verge of violating in the context of cultural reality a fundamental aspect of man's uniqueness, a transgression for which we have heretofore prepared by a violation in the abstract and intellectual domain?

If this is not true, then we must explain why it is that with disconcerting frequency these predictions go awry and even the mass behaves in an unexpected manner. In response to this novel and untoward outcome, however, the model is only revised, made more exact, the internal structure altered. But the model is rarely discarded. For within this paradigm of human behavior is a vision of rationality, of the rule of law. The assumption must be reinforced that in living processes also, nature is homogeneous. Its invariants and constancies apply to the entire range of phenomena, always and everywhere.

3

One of the most paradoxical examples of the pervasive influence of this theory, and perhaps the only large-scaled social exemplification of the attempt to put into practice even a modified version of the biological world view, is given in the recent history of American public education. The progressive movement in education arose and came to fruition at about the turn of the century, a time not only of great social change but of important intellectual transitions, especially as it saw the evolutionary theory begin to affect thinking in the different biological disciplines.

In education, John Dewey, Edward L. Thorndike, and William H. Kilpatrick were preeminent. Thorndike was in effect the founder of educational psychology, one of the first to attempt to put the scientific method into use in educational practice. In his so-called "connectionism" he was deeply influenced by the stimulus-response bond and firmly committed to an observationalist approach to data, witness his influence on testing and I.Q. Attempts to understand the nature of learning and the learner were pursued in the belief that these matters were amenable to external analyses and subject to intellectual and practical control.[36]

Kilpatrick, though a student of Dewey's and perhaps the most well-known purveyor and adaptor of Dewey's philosophical view as they applied to school practice, was influenced even more than Dewey by the biological-behavioral frame he received from Thorndike.[37] And it was through the great pedagogical impact of Kilpatrick that over a generation of teachers was imbued with the biosocial adaptive view of education. It was Dewey himself, without doubt the most influential of all the educational innovators and the symbol of this progressive era, who provided the paradoxical twist in the movement. Without Dewey's transmutation of the raw behaviorism that inhered in both Thorndike's and Kilpatrick's outlooks, the progressive movement in education would have been beset much earlier than it was by the internal contradictions which eventually vitiated it. Dewey altered his originally orthodox Protestant and Hegelian attitudes soon after he studied the biological and evolutionary views espoused in William James's *Principles of Psychology* (1890).[38] (His earlier exposure at Johns Hopkins to the scientific and biological views of G. Stanley Hall had left him untouched.)

On the other hand, it was not too long after this shift and well before the heyday of behaviorism that Dewey wrote an article entitled "The Reflex Arc Concept in Psychology" (1896), which set forth in clear and incisive terms his objection to the simple reduction of human thought to the sequence of stimulus-reflex responses.[39] It should be stated that at that time both Thorndike and Kilpatrick were far to the right of Thorndike's later extreme reductionism—his residual mentalism, epitomized in the aphorism that "animals think things, while humans think about things." Nevertheless in their attempts to objectify human behavior and subject it to scientific analysis and some measure of control, their ideas became permeated by the stimulus-reflex hypothesis, which in the human societal and educational context necessitated considerable theoretical ingenuity.

Even further from the traditional manipulative ideals of behaviorism was the movement's genuine commitment to social amelioration and liberalism, greater freedom for both teacher and student, and a deep concern for the uniqueness of the individual.[40] With the advantage of hindsight, one can say that this educational movement was the antithesis of any program that emphasized determinism, manipulation, extrinsic motivation, and mechanical uniformity. Rather the movement bore the stamp of an intellectual tradition that boldly asserted man's unique creative qualities. It was characterized by its openness to the new and innovative and its acceptance of social and political movements based upon an imaginative realization of human needs rather than on the satisfaction of fixed and anterior drives.[41] It was

Dewey's instrumentalism that philosophically carried these varied and seemingly antithetical strands into the educational movement. Progressivism in education succeeded because for a short period it seemed to bridge these opposed qualities. The reforms in education and society it sought to effect seemed to be derived from these intellectual tenets. Perhaps it was this basic intellectual incompatibility between biological adaptation and social freedom that undermined and finally reduced the impact of this tradition both in the social and political spheres as well as in the educational arena.

Dewey's Darwinism was rooted in his view that nature, whether physical, biological, or social, is in a state of continuing flux and change.[42] The evolutionary process becomes an ongoing process irrespective of the particular forms that are created as an outgrowth of its movement. Dewey postulated that man's prime need is to develop a social atmosphere that will put human institutions in a state of dynamic equilibrium with the environment. The method of thought enabling man to create this changing social stasis Dewey called intelligence. Putting the method of intelligence to work in the education of the young, in teaching the child to learn how to learn, or how to apply his intelligence to a changing world at all stages in his own development, is in effect the scientific method.[43] Rather than focus on the traditional subject matters of education, Dewey focused on the creative source of man's new subject matter—scientific intelligence. Dewey admitted drives to his psychosocial schematology. But they were undifferentiated instinctually and infinitely malleable into societal behaviors. Implicit in man's renewable creative flow of energy was his capacity to direct thought into scientific modalities which acted as a critical force within any cultural setting. Dewey thought of scientific intelligence as a supracultural objectivizing element in human nature.

The role of thought thus became critical, practical, and utilitarian. In education the theme was likewise practical—life adjustment or social adjustment. In Dewey's conceptualization the very process of catching up with a changing world necessitated radical alterations in any present structure of society.[44] But in the realization of these philosophical ideals as utilized by educational administrators Dewey's program became a rationale for imposing a kind of social quietism on the student to mesh him with the status quo. Education to Dewey was a "third thing" mediating between the external changing environment, both natural and social, and the particular institutions of society, usually conservative and somewhat repressive. Education, then, using the medium of intelligence to foster intellectual control, became the social correlate of the fang or the claw.[45] The kinds of educational subject matters that constituted the goal of a universal social intelligence were broadly conceived by Dewey. His belief in the practicality of the humanities and the arts and his continuing concern with them are testimony to his breadth of understanding with regard to the diversity of interests in human thought.[46] But the basis of his concern was always whether these aspects of human culture effected a better adaptation in the flux of the evolutionary process.

Thus, education in America, from the efficiency experts administering public education to the problem-solving and project methods curricula that were developed for the schools, was conceptualized in a manner completely reconcilable with the evolutionary point of view. One is taught to learn how to learn, to think critically, to search for an intellectual consensus, to grasp what is public knowledge. This was a time of straightened economic circumstances for the masses, therefore the orientation of education was ever more practical and vocational. One did not study a subject matter or a discipline for its own sake. Subject matter had to focus on life and inevitably on the current exigencies which social existence presented.

Rather than effect a criticism of social experience, American education began to reflect the more orthodox function of adapting to contemporary society. Independence was increasingly vitiated. The eventual demise of Dewey's influence coincided with the complete and conscious intellectual absorption of the American school into its traditional role of reflecting and enhancing the informal value system of our society. The cause of this disheartening denouement perhaps lay as much in the nature of the philosophy which had directed its changing character as in the failure of institutions and individuals to live up to an ideal.

By taking a historical perspective of the rise and fall of this important social and educational movement we can understand the direction of that series of events. Darwin's *Origin of Species* represented the introduction into modern thought of one of those rare but crucial archetypal ideas in the history of culture. It had the empirical and theoretical authority to sweep away the vestiges of the old and flaccid metaphysical and theological integrations to make way for an entirely new basis for reconsidering the meaning of experience. The new *a priori* invariances for thought which it set forth stimulated new disciplines, new theories, and a revision of some of our most basic attitudes toward man and life. The results have been deep and far-reaching in fields of study far removed from the particular structural confines of Darwinian evolution. Like all new theories its greatest contribution came in its initial impact; it let a breath of fresh air into science. A large range of disciplines profited by the rapid conceptual advances fostered by its novel insights.

But like all theories it has aged. What can be called normal scientific investigation has run its course in explaining all the possible implications and deductions that can be formed from its perspective. While Darwinism has by no means been modified to alter its basic claim concerning the function of natural selection in the evolutionary process, the process itself is seen to be of great complexity. There are still too many subtle questions to be resolved to give an objective observer the feeling that the evolutionary process is a kind of omnipotent juggernaut which drives irrevocably ahead trying to meet its timetable in history.

Evolutionary theory in its recent decentralization of focus and research is far more interesting to us today in the numerous specific problems it has dredged up. While because of its almost mechanical and deterministic vision of life on earth it lacks the excitement and exhilaration in terms of philosophical expectations, it still provides an intellectual center from which both the life sciences and the cultural disciplines must work. Scholars today are less sanguine about claims for absolute objectivity in any of the allied fields. And if these disciplines cannot ape the predictive capacities of physics, at least the latter subject matter has paused momentarily in its hurtling advance since the revolutionary days of the quantum. In practically every discipline which touches upon the evolutionary theorem (which yet remains fruitful as a maxim for thought), the simplistic extrapolations of the adaptational and survivalistic model have been subject to serious qualification and reconsideration. Diverse postulations now receive an attentive hearing. The ground may be readied for another look and a new paradigm.

4

It is interesting to note that the continental climate of opinion was so constituted that it absorbed the Darwinian doctrine with somewhat less of a convulsion than did the English-speaking nations. Certainly the impact on religion was a strong factor in stimulating the subsequent debates. That Darwin was an Englishman would also explain the tremors which took place at home. Add to this the fact that the Continent was still a refuge from scientism and that amidst the metaphysical excesses in philosophy there were strong theoretical traditions in the various humanistic disciplines which buffered the Darwinian impact.

Cassirer was familiar with the various philosophical issues with regard to the structure, function, and historical development of life that were stimulated by Darwinian teachings. In *Das Erkenntnisproblem* Cassirer dealt in great detail with the biological world view.[47] It is interesting to note that he discussed this view in terms of the inner dialectic of the discipline and not in terms of its possible extrapolation to undergird theories of man or culture.

Thus the significance of a theory of symbolic thought, left

without consideration in the third volume of *The Philosophy of Symbolic Forms*, had not as yet been joined to the issue of evolutionary biology. It is possible however that this lack of attention to his own systematic philosophical concerns was due to his methodological restraint in dealing with the subject matter of biological theory solely in terms of the contextual issues attendant to their discussion. It was not until he arrived in the United States and wrote his *An Essay on Man* that he faced the problem of knowledge and a theory of human nature directly. It now revealed the essential and important logical questions heretofore only implied in his philosophical writings.

But Cassirer reverted to his more conservative epistemological stance in his last work, *The Myth of the State*. He still charges Freud with being a reductionist, yet does not address himself to the cultural sources for the empirical content of Freud's biological reductionism. We can note here that it was not accidental that Freud's influence flowed westward and that his ideas were well-received in those centers of thought that were already prepared by evolutionary theory for his radical theoretical revision of the dynamics of the psyche. However, Freud, like Dewey, was too far-seeing to subscribe to any simple reductionist scheme and, as Susanne Langer has noted, there is much in Freud's view of the unconscious to support and add to Cassirer's conception of the mythological mind.[48]

Perhaps it was his continental education which endowed Freud's picture of man with such philosophical depth. There is rarely the myopic disregard of the total range of human thought that is evident in so much behavioral psychology. That his process of enquiry was pursued into the depths of the psyche, that he was constantly revising his "metaphor" of mind and widening his theoretical vision to account for broader rational and societal considerations, attest to the qualitative differences between Freud's vision of man and society as compared with, for example, B. F. Skinner's. Also his recognition of the role of reason and ego in bringing man into social accord with his fellows and his acknowledgement of the reality and importance of such non-biological qualities as dreams, gestures, ritual, and religion attest to Freud's realization that evidence contrary to a simple genetic and biological view must nevertheless be accommodated philosophically to a more flexible evocation of the nature of man.[49]

The supporters of ego-psychology (a recent outgrowth of Freudian tradition)—Erik Erikson, Heinz Hartmann and Heinz Werner—reflect a greater appreciation of the role of reason, with all due consideration of the inner dynamics of thought as stabilizing and integrating factors, in the accommodation of man to man.[50] The trend within the mainstream of Freudian psychology seems increasingly to be to emphasize the role of reason as a means of self-maintenance for the fulfillment of man's inherent potentialities, though earlier philosophical deviants from Freudianism argued that the orthodox conceptual structure, with its reliance on drives and its conception of human development in terms of deterministic and unilinear ontogenetic processes, had to be abandoned. In the writings of these so-called neo-Freudians and a growing number of renegades from behaviorism we note a persisting attempt to understand man's thought and action in terms of its autonomy from predeterminant conditioning.[51] Behavior in this view is intentional, purposive, and goal-directed; it is self-maintained through a rational and cognitive integration of societal meanings. In short, the more man becomes truly one with his own nature the more he acts in a manner at variance with what the biogenetic psychologists might predict.

In recent years greater respect and authority have accrued to a school of psychology whose Kantian roots are explicit and which paralleled the neo-Kantian tradition in the physical sciences. Gestalt psychology has always been committed to a position that the products of thought and perception are grounded in the laws of thought and not in the given of external phenomena or in anterior biological drives and needs. Theories concentrating on themes such as the structuring of sensory experience, objectivization in thought through symbol, and esthetic creativity have argued for the autonomy of thought as against the supposition that behavior is dominated by insulated drives or rigid, ontogenetically rooted determinants.

This heterodox tradition has been supplemented by a rich store of empirical evidence from the stronghold of biological determinism—developmental psychology. No doubt many of the motifs of Freud—repression, fixation, regression, transference—have been of great heuristic value. However, many of these concepts have been of significance in understanding, not the normal child and his mental growth and development, but the psychologically abnormal child.

The Gestalt model has, however, predominantly concerned itself with normal growth and thought. Thus the work of Bühler, Stern, Piaget, and Vygotsky has been concentrated on the intellectual development of the normal child as he masters the world conceptually and linguistically.[52] Here the process is dominated by such theories as self-actualization, internal speech, categories of thought, all of which echo the assumption that a human being has the inner potential and need to organize experience in terms of objectification and symbolic action.

The interaction of linguistics and anthropology has resulted in diverse philosophical trends because of the inevitable qualitative and cultural considerations that are brought to bear on these fields of study. The work of the nineteenth-century continental linguists—from Humboldt's philosophical grammar to Saussure and Trubetzkoy's work in establishing structural linguistics— represent a variety of empirical, descriptive, and theoretical concerns. The unique problems encountered in the study of language immunized this discipline from the early metaphysical and theological involvements of philosophers and psychologists. Thus while the advent of Darwin shook so many disciplines into their own forms of behaviorism and biologism, the field of linguistics developed a more pluralistic quality.

There have been, of course, many recent attempts to discover universal grammars, of finding in the external syntactical and grammatical relations of language a logical system that might be mechanically fed into a machine and translated into other languages. There have also been attempts based on the assumption of the existence of logical universals in language to reproduce mechanically the phonemic structure of a language. But to balance these attempts, there have been thinkers, as we have noted earlier, such as Edward Sapir, Benjamin Lee Whorf, and Karl Lashley, who were aware of the richer and more unconscious sources of language use. In recent years, however, infatuation with mechanical, mathematical, and behavioral approaches has once more waned as hopes for understanding human language through replication and logical simulation have been disappointed.[53]

The younger generation, represented by the linguist Noam Chomsky, now reiterates many of the traditional views concerning the internal structural unity of language systems, their conventional and culturally rooted syntactic and semantic characteristics and thus their unique status within the spectrum of general animal behavior. Indeed Chomsky's philosophical view of language—a return to the older idealist and rationalist position—is based precisely on the inability of the behavioral and taxonomic models of language acquisition to reveal the essence of language function.[54]

Several qualities of man's use of language illustrate the existence of internal thought processes and structural principles which do not mirror the world of external experience, whether interpreted through societal interactions of the individual or his perceptual awareness of the state of things.[55] Examples of the existence of the inner structure of thought may be seen in the striving of children to express their ideas, to communicate in obviously garbled or telegraphed syntax when their actions indicate an awareness that far outpaces their language abilities. Another exemplification of language's deeper roots in our mental structure can be seen in man's capacity to understand even when the sensuous message is so distorted that a machine would identify the message as noise.

In fact, man's divergence from the so-called empiricistic or mechanical view of language is made more significant by its dissimilarity to animal communication.[56] The closest man comes to animal communication is in the system of gestures—for instance, in the visual hand movements we make in directing someone to park a car. Otherwise human language as compared to animal signs is a totally new biological phenomenon. Nowhere in animal life or in machines do we find (a) the possibility for generating

an infinite number of new sentences from a relatively small repertoire of linguistic materials, words, and sounds (generative grammar); (b) the ability to respond freely and unpredictably to
similar stimuli, linguistic or other (man's speech in any situation
is indeterminable; the level of deviation here also approaches an
infinite number of possibilities); (c) the ability to disregard the
particular external and formal characteristics of grammar and
syntax and to respond linguistically to the situation's inner meaning in an appropriate manner. Thus it is the inner meaning of
words that moves man to act rather than the grammar and syntax
of the words.

These factors, among others, cause Chomsky to wonder about
the validity of evolution, about man's real nature and about his
place in the biosocial world. Although a student of Cassirer's
thought, Chomsky does not look upon this surface lack of amenability of evolutionary theory and linguistics as a challenge to his
own Kantian and Cartesian conclusions.[57] The theoretical and
empirical powers of this new view of language are perhaps too
vivid as yet to bring forth the kind of concern that Cassirer manifested at the end of his career. Nor can we make any judgment
beyond reiterating what all contemporary idealistic views of man
and culture must conclude—that the theory of evolution as pres-
ently constituted does stand as an ominous barrier to every extension and generalization of these views.

We might note that, in anthropology also, investigators such
as Boas, Kroeber, and Malinowski have stayed far away from the
biological model. Indeed the cultural evidence was a consistent
intellectual barrier to such tempting theoretical extrapolations.
Models of thought in this field have consistently remained functional and descriptive rather than global and reductive.

5

There have been many examples in the Western intellectual
tradition of the interaction of important ideas and governing

social conditions which have given rise to an enduring *Weltan-schauung*, an epoch defined in terms of philosophy and life style. Christianity shared its medieval era with agricultural feudalism. The age of reason began with Galileo and Newton and was infiltrated by laissez-faire mercantilism.

The nineteenth century began with the infinite vistas of mechanical theory and industrialism. Later the writings of Darwin, Marx, and Freud gave indirect support for a burst of economic and technological growth that soon changed the face of the West. These intellectuals meant more to the larger social community than the specific contributions they made to their respective disciplines might lead one to believe.

At the turn of the century, when Ernst Cassirer began his very specialized studies in Marburg with Hermann Cohen, he could not know in which direction his studies would ultimately lead him. The aura of Kant and the excitement of studying philosophy with the foremost of the contemporary neo-Kantians was rationale enough for this exceptionally talented young scholar. But in following the Kantian path he discovered that it led constantly outward toward newer and wider intellectual vistas. Eventually he arrived at the most generic expression of the original epistemological problem, the question of human nature.

But in locating this ultimate radiation of the problem of knowledge and then of thought, we cannot presume thereby that the issue will remain wholly in an abstract and theoretical domain. Ideas have wider consequences. Every day brings the necessity for decision-making in crucial areas, such as politics, education, and social policy. With each decision comes the necessity to choose between alternatives. A basis for making decisions is necessary. Even the kinds of piecemeal engineering advocated by Karl Popper demanded at least short-term criteria. The grounds for choice always point to larger philosophical considerations if the vision is there to pursue them. But even if well-considered actions are not taken with regard to specific situations, it is wrong to think that the cumulative decisions made in a society will be either random or limply acquiesced to on the basis of actions in the immediate past. There are larger patterns and directions to decision-making seen in terms of the overall trend of events. In like manner, what has occurred in the West in the last century and a half is a pattern of social decision-making that has been under-girded by certain archetypal attitudes toward human nature.

During this period there has been an accelerating technological assault on our planet in the name of lifting up the standard of living of the masses, freeing them from their bondage to the land. The material wealth has flown out of the land and carried with it the incubus of want from a large portion of mankind. Where-ever one sees the beginnings of this process take place, one notes only that it is given universal assent.

The humanist protests against the profanation of our environ-ment, the rat-ridden slums, the grimy factories, the anonymous criminality, and the horrors of technological warfare. Could there have been another way for modern society to protect us against the random vicissitudes of nature?

The material view of the universe given in physical science as well as the picture of man given in Darwinism has been critical as a pillar of support for the events that have taken place on the social level. It cannot but have been that much easier to warp the cultural, psychological, and now the ecological balances that had been established over the millenia by means of this now per-suasive argument: what we do in terms of material and economic growth fulfills a basic and ineluctable drive of man. The long history of man has presumably been a continuing quest to adapt to the environment by subjecting it to his will. How supportive to this view have been the writings of Darwin on evolution, Marx on economics and material nature, Freud on man's inner drives? This drive to power has had its contemporary realization in our technological society.

More recently, events have conspired to argue that this is a potentially defective vision. The one-sidedness of our preoccupa-tions is increasingly evident as we must begin to repay our debts to our culture and to nature. To act on this threat with intelligence, some would argue, a new intellectual integration now seems necessary.

One wonders if the breadth of perspective in the Kantian tradition might offer some contemporary light. Still, a key theoretical gap remains. The roots of symbolic thought need to be established more deeply in the evolutionary tradition. In the next chapter we will set forth certain factual and theoretical considerations from the recent literature on evolution that support Cassirer's conception of human nature.

8

Man as Symbolic Animal:

An Evolutionary Interpretation

CASSIRER LEFT US WITH A DIFFICULT THEORETICAL dilemma: the existence of two contrasting and opposed views of human nature. This problem is as intellectually significant now as it was then. Yet in terms of the impact of Cassirer's writings on the philosophical world it led to negligible efforts. The causes for this were varied. For the most part the biological paradigm was still being actively mined in a number of disciplines. In addition, Cassirer's historico-critical approach and his use of highly technical disciplines to support his views did not result in as "revolutionary" philosophical claims as did pragmatism and logical empiricism.

One cannot claim that the latter third of the twentieth century is a more propitious time for a reevaluation of the impact of neo-Kantianism than was the middle of the century. However, the confidence that was invested in the regnant philosophical and scientific models has been somewhat dissipated. Instead of the energetic materialism and determinism which gave rise to an entire generation of electronic and mechanical automata—the cybernetic era—with its confident extrapolations into the social and cultural arena, a general dissatisfaction with science and technology has now set in.

For a while existentialism and neo-orthodox religious philosophies provided an alternative. Those too are now for the most part ineffective in giving any intellectual coherence and breadth to our philosophical vision. It is difficult to accept as fact that the door is closed to a broader, more encompassing perspective of man as a natural, secular, biocultural creature. Man certainly will not rest until he finds coherence and meaning in the whole fabric of experience.

It is here that evolutionary theory continues to provide the key to the grasping of the role and significance of man's place in nature. It is the seminal theory of the biosocial world and at the same time is very different in structure than the theories of physics and the inorganic realm. Here the element of historical change becomes crucial. We must strive to keep whole the principles which undergird this process of change. The basic principle of rational thought, the maxim of the "continuity of nature," demands this of us.

What is the alternative to the theory of evolutionary biology? How can we unravel the various dilemmas which obtrude from an examination of human behavior, culture, and knowledge? If we put aside our search into evolutionary theory as part of an explanation of structure and function, are we not then left with some mysterious inner propensity of man? Irrationalities come in various forms. They are not necessarily intellectual or philosophical. They express themselves in drug cults, in pseudo-religious communalism, in disaffected "flower" children. But is their existence not due ultimately to the disintegration of the fabric of meaning in a society? When the intellectual strands are so tangled that the direction of social and scientific study is without a focus, it is no wonder the culture begins to lose its élan.

True, there are great difficulties in conceiving of symbolic behavior as a product of the evolutionary process. Perhaps we could accept this as an insuperable difficulty and turn instead to descriptions of man in culture in functional and structural terms. We could, as Cassirer did at first, restrict ourselves to a phenomenological analysis of the various kinds of knowledge in the plurality of cultural contexts. The symbolic theory of culture and knowledge could be fulfilled in part by restricting our study to an examination of the differing psychological orientations of each discipline and by describing each particular internal logical structure. In this way we would avoid the need for global explanations of human nature and culture. But again, it is hardly likely that this state of theoretical explanation will prove to be a necessary or sufficient condition for the promotion of a state of intellectual satisfaction.

Man is enough of a mystery that we can investigate him in terms of more limited theoretical constructs. We can study social structure, the relationship of creativity and esthetics to art, even the phenomenology and internal logical structure of the kinds of knowing. The list might and does go on *ad infinitum*. But as Cassirer noted there are two pervasive themes in the evolution of scientific knowledge—ideality and universality. And certainly the latter condition remains unsatisfied in terms of the historical dimension of man as long as the evolutionary puzzle exists. The succession of cultures, the development of the arts and sciences, political and religious institutional forms float in an intellectual limbo when we cannot drive our causal regularities deep into the biological stratum in search of the source of such human behavior.

The issue of a choice between either the behaviorist or the symbolic view is not one of gross facts. The evidence in those disciplines that are associated with evolutionary theory is vast. One cannot expect that a single newly discovered factual speck will render one or the other position as patently false. Where the issue will probably be decided will be in the area of interpretation, where the evidence will be fitted together in a newly coherent pattern. We will then have a shift of opinion to a new center. Either a new way of interpreting the biological-behavioral constants will help clarify the intractable cultural elements associated thereby, else there will be a modification of the seemingly univocal adaptational interpretation of animal behavior, allowing man a legitimate cultural entree into the vast stretch of evolutionary events. The old, whatever view it represents, will die slowly; its base of intellectual support will gradually erode and wither. Certainly Cassirer's neo-Kantian epistemological and cultural position is valuable, at least for stimulating further research and enquiry.

Thus we ought not conceive of the Darwinian laws as metaphysical absolutes. We can accept the vast accumulation of empirical materials supporting the fact of evolution without committing ourselves to a belief in the permanence of the particular intermediate conceptual constructs and mechanisms now used to explain the general process.

This commitment to a simple cause and effect delineation of evolutionary theory has been the source of the various biologically oriented social and behavioral theorists. They have made a mental commitment to one interpretation of the theory, an interpretation which superficially appears to be rigidly embedded in the nexus of facts. As a result of this narrowness, they have had to deny some apparent facts of experience in favor of the supposed truth of their abstractions.

But if we are to support a neo-Kantian view of man, thought, and knowledge, we cannot merely raise these methodological caveats. For this position of critical-idealism to achieve acceptance as anything more than an interesting and plausible hypothesis, it should connect with a theory of evolution. Man must be shown as a cultural and biological being not bifurcated conceptually into two incompatible "natures." To satisfy such an intellectual demand, as Cassirer demonstrated numerous times in his own writings, one must leave abstract and methodological considerations for the moment and turn for the answers to the empirical data compiled by the scientists.

The demand for historical continuity in nature will therefore lead to a demand for structural continuity. We will ask not only how man evolved from his simian progenitors but how the structural differences in morphology and behavior came to be associated with the gross alterations in the intentional pursuits of man. An explanation of symbolic thought from the standpoint of evolutionary theory will not necessarily require the historical tracing and delineation of these morphological differences. The work of researchers in aphasia and neurology has progressed to the point that we are gaining insights into the novelties and uniquenesses of the human brain. The issue that needs explication is how man's behavior was shifted from a system of crudely adaptive reactions to a system of symbol behavior, how a completely natural morphological change resulted in behavior which from the standpoint of the traditional requirements of natural selection was only incidentally adaptive in function.

One is not prejudicing the outcome by going to the problem of evolution to search for a structure of ideas compatible with a neo-Kantian theory of knowledge and a symbolic interpretation of man and culture. In order to achieve objectivity, one need not engage in research while wearing a philosophical blindfold. Our only methodological requirement is that the investigator should not consciously ignore or turn away from evidence, either factual or conceptual, that might disconfirm his quest.

1

Let us turn now to the problem of evolutionary theory as it concerns the coming of man. We will attempt to establish that Cassirer's symbolic view of cultural thought—implicitly, his interpretation of scientific knowledge in terms of intrinsic cognitive drives and of purposeful and intentional behavior—is compatible with a theory of evolution that views man as part of that same natural process which has produced all organic forms.

Some of the more important aspects of recent genetic interpretations of evolution are as follows: all interbreeding, panmictic species share a gene pool within which inevitably occurs a fairly constant number of small and random mutations.[1] These mutations can occur in all morphological, behavioral, and physiological directions and can spread throughout the gene pool, depending on their recessiveness or dominance and on conditions internal to the character of species, e.g., distribution, length of generations, rates of reproduction, etc. The rate and character of these mutations, however, have nothing to do with the exigencies of external events. The fact that every species is in a state of continuous interaction with a dynamic and changing environment casts the mutant genes in a crucial role. Inevitably there will be an accentuation of the existing state of ecological equilibrium or disequilibrium in the species. In concrete terms the effect of this dynamic interaction would be either to buttress the adaptiveness of the population, thus maintaining a favorable selective ratio, or, conversely, to cause disequilibrium and disadvantage to the species in reproductive terms in the face of a now more hostile physical environment or in the face of competitors who might now be increasing their ecological advantage.

The balance of nature is so delicate that in most circumstances a slight disadvantage in bringing the normal number of young to maturity would lead to a relatively rapid extinction of the species. The process of natural selection would have then been invoked to effect its ruthless yet random results on those species not able to adapt sufficiently to maintain themselves against the intense dynamics of the evolutionary process. This general schematism applies as much to man's ancestors as it does to other living creatures. It is when we approach a specific analysis of those features of human morphology for which we can find an adaptive purpose that serious difficulties begin to intrude.

Today we recognize several events in man's development as being crucial to his humanization—the abandonment of a prehensile foot, preceded by the displacement of brachiation as a means of locomotion with the upright gait, and the development of the opposable thumb. But the most important development and the key for utilizing these changes was the growth of the brain.[2] If it is indeed possible to interpret the growth of the brain to its present size and structure as a process strictly within the confines of the genetic theory of natural selection—that is, that at each stage of its development its functioning meshed closely with the selective demands of the environment—then we must abandon our interpretation. It is inconceivable that an organ whose evolution was carried forward for so many eons because of its concrete and practical utility for the survival of the species could suddenly alter its basic character in historic times to function in a manner that would make these age-old adaptations irrelevant.

Is it possible that the brain did not serve an adaptive purpose, at least at the stage in its development when it gave rise to homo sapiens? Also, could the process of natural selection itself have been inoperative, to allow for the growth of the brain regardless of the imperatives of survival and selection, and thus to reach an adaptive level in which new and unique selective criteria could have become established, i.e., culture?

For the most part we can interpret the evolution of the infra-human primates as being compatible with the orthodox evolutionary schematism. Anthropologists have traced man's probable ancestors to Parapithecus and Propliopithecus of the Oligocene period, and Dryopithecus and Proconsul of the Miocene period, both rather undifferentiated progenitors of the *Hominidae* and existing *Pongidae* (anthropoid) stocks.[3] Within the range of their special ecological requirements, these were both well-distributed and flourishing forms.

These creatures had already benefited from the gradual growth of the brain and the concomitant reduction of the snout, a process which had been occurring since their lemuroid and tarsioid ancestors had taken to the trees some thirty to forty million years earlier. In addition, the development of binocular vision, the reduction in the number of offspring born to the female, the increased care expended on the few young, and a general selective shaping of these arboreal creatures toward a sensitive awareness to the dangers in the environment and an ability to avoid direct physical confrontation with possible enemies all contributed in a positive way to the correlation of intelligence with selective advantage.

Now, if we could trace this steady interaction of genetic change with adaptive advantage from the Miocene through the late Pleistocene (when homo sapiens appears), we would have no basis for argument with the strict selectivists. But the evidence from a number of allied fields adds elements which cannot be absorbed into a view which argues for a smooth and deterministic interaction of genetic mutations for brain size being correlated with selective advantage enduring over a continuous span of time. The issue itself does not bear on the concrete facts of human evolution. But it raises serious theoretical qualifications as to the assumption of a unilinear selective process.

Let us first examine the basic facts that are available on human evolution. The beginning of the Miocene, roughly twelve million years ago, was marked by great ecological changes. It also coincided with a blurring of the fossil record.[4] Sometime during this period the lines of man and ape must have separated. By the late Pliocene or early Pleistocene (the ice age) roughly two million to 500 thousand years ago there was a wide variety of proto-hominid creatures. The most primitive, Australopithecus, who was capable of acquiring a rudimentary material culture, had at this time a brain capacity of slightly more than the 600 cc. of the contemporary chimpanzee.[5] A little later and perhaps even overlapping these creatures came the pithecanthropoids (Java and Peking man). These creatures were far more advanced, with a brain capacity of 900 cc. and a much wider variety of cultural achievements—fire, tools, etc. Sometime during the middle to late Pleistocene, from 250 thousand to 50 thousand years ago, we encounter fossils of the neanderthaloids. The neanderthaloids are closest to modern man morphologically, yet for the most part not on the same line of development. These still primitive creatures, while much advanced over their precursors and having a brain that averaged over 1540 cc. (larger than modern man's), were still coarse prognathous and apelike creatures by comparison with homo sapiens.[6]

There is no record of the Neanderthals after the end of the Pleistocene, about 50 thousand to 25 thousand years ago.[7] According to the fossil record their disappearance coincides with the coming of homo sapiens. There are some inconclusive data which indicate the possible existence of a primitive but distinctly sapient creature (as distinct from homo neanderthalis) some 200 thousand to 80 thousand years ago.[8] The drift of the evidence, however, points to the fact that man burst upon the evolutionary scene with a bulging cerebral cortex, hairless, without instincts and sexual periodicity, armed with language, art, technology, and a cultural ferocity that may have annihilated the neanderthaloids at their every point of contact.[9]

2

Man's sudden arrival presents to us yet another evolutionary puzzle that has not been clearly answered. If we examine the paleontological record alone, there are two important and crucial facts which strike the observer and two important theoretical questions arise from these facts. Fact one: Rather than a clear-cut line of descent from such primitive yet unspecialized apes as Dryopithecus or Oreopithecus, we have not only the radical branching and splitting of the *Hominoidea* (super-family) into two distinct families, *Pongidae* and *Hominidae*, but we have the latter splitting again into distinct genera, the *Australopithecinae*, having a wide variety of forms, and the other *Hominidae*, which includes the various pithecanthropoids in one genus and Neanderthal (many types) and sapient man in the other (Homo). This seeming fragmentation of a formerly stable and continuous ancient line of descent has moved several commentators to suggest the development of genetic instability as the evolutionary line surged forward out of Pliocene into the Pleistocene.[10]

Fact two: In addition to the fact that the great number of types which came forward presumably from but a few ancestral forms in the Miocene, we have the puzzling reality that this splitting into new taxonomic categories took place in a relatively short period of time. We have a transition in forms from the Australopithecus with 600 cc. to *Homo neanderthalis* with 1500 cc. in roughly 500 thousand to one million years.[11] There is other evidence that would put the length of the entire Pleistocene age at

300 thousand years, which would make the development of man (his "hominization")—even if Australopithecus could be placed back into the Pliocene—a fantastically rapid evolutionary affair.[12]

The first problem which arises concerns the propriety of using the theory of natural selection, with its basis in genetic theory, to explain this very rapid evolutionary process. Through genetic theory we postulate that mutations take place at a steady rate, completely uninfluenced by environmental factors.[13] We also know that evolutionary change proceeds at varying rates. The paleontologist George Gaylord Simpson has schematized the varying rates of change in the history of life as ranging from bradytelic to horotelic to tachytelic (slow, average, and fast).[14] This would imply that the differentiating factor in the evolutionary process is located in the external environment, which acts on the available store of genetic variability. Obviously in certain situations the selective process will act rigorously to accentuate differences and enact great changes. At other times it will allow variation to "pile up," to be internalized within the taxonomic category, usually the species, leading to subspecies and races, without propelling these changes into more dynamic interactions with the environment, which would cause them to generate wider taxonomic differentiation.

The major need is to investigate the plausibility of the claim that evolutionary progress is procured by the action of natural selection on the steady accumulation of minute genetic variations, which are themselves heterogeneous, and that it takes place in all directions—morphological, physiological, and ethological (behavioral). Our concrete problems concern the applicability of the theory to the issue of the growth of the brain. What must be accomplished is the correlation of a number of conceptual factors: a mathematically determinate theory of mutational change as well as a determinable theory of the distribution of mutations through the species, given considerations of dispersion of interbreeding races, size of population, rate of reproduction, generational rhythms, etc. In addition one must consider the very indeterminate factor of relative rates of positive as against negative mutations over the total range of variations. To add to the difficulty of using the theory of natural selection to explain this rapid brain growth, it has been postulated that homo sapiens has been genetically stable during the past 25 thousand to 50 thousand years.[15]

Loren Eisley concluded that, on the basis of the difficulties of squaring the paleontological evidence with the genetic theory of natural selection, the morphological development of homo sapiens is at present unsubsumable to the traditional postulations of the theory.[16] Why, he asks, should nature have set off such a radical trend toward increased brain capacity from 600 cc. to the 1350 cc. average of man, when selectively 900 cc. would certainly have done as well.[17] How could creatures so well adapted in the Miocene age possibly need brains of sapient structure and capacity to survive conditions of the Pleistocene, even considering the grossest ecological changes that took place in the intervening time.

Eisley adds another caution to the selectionist views on the adaptive value of brain size. He cites the mysterious case of Boskop man, a true homo sapiens with a brain capacity of 1600 cc. and a remarkably sapient, foetalized visage. Presumably this creature, far more than ordinarily endowed with potential for intellectual and cultural adaptation, was in some manner bypassed by evolution. (He became extinct in the late Pleistocene.) Boskop man, a possible great uncle of our present day Bushman, may have been the symbolic animal par excellent.[18] His mysterious demise makes us wonder even more at the miraculous extrusion of our own species from the seminal fires of this "macro-evolutionary" process.

The second theoretical difficulty is symptomatic of a more general issue within evolutionary theory and is addressed to man only to the extent that human evolution partakes of the larger problem. It is concerned with the taxonomic direction of evolutionary change as it can now be subsumed to the genetic theory of natural selection, which is that aspect of evolutionary science that most closely approaches the deductive achievements of the physical sciences. By and large selective processes act on the gene pool in such a manner as to lead downward in the taxonomic scale of changes. Thus in an examination of such factors as mutation rates, kinds of mutations, and rates of transmission of mutational change, in the context of the selective processes of a changing environment, one obtains an unexpected result. Through natural selection evolution, instead of producing larger morphological and thus greater taxonomic changes, produces smaller categories. Natural selection acts to subdue the rate of evolutionary progress to the point where racial changes are the norm, speciational changes relatively rare, and new generic creations seemingly impossible. And yet the evidence from the fossil record hardly confirms the supposed direction that this presumably deductive and scientific theory would indicate.

Perhaps the issue is best illustrated by the mathematical treatment given it by the geneticist Sewell Wright. His work underlines the current perplexing scientific state of the traditional Darwinian assumptions about evolution. On the basis of a study of the rate and character of genetic alterations, he has concluded that all creatures are engaged in a battle to maintain themselves against genetic deterioration and fragmentation. The mutation rate, rather than providing the direction and impetus for evolution, leaves the creatures in a drawn-out competition with the environment. He expresses this puzzle as follows: "While the process as a whole has the aspect of a struggle of the species to hold its own in the face of a continually deteriorating environment, rather than of evolutionary advance, there can be no doubt that a large part, perhaps the major portion of evolutionary change is of this character [advance]."[19]

Herman Muller has buttressed this view by showing that the frequency of harmful mutations has caused the rate of mutation itself to drop.[20] On the basis of these authorities, at least from the standpoint of the orthodox neo-Darwinian view, we can conclude that subspecies are rarely incipient species and that the multiplication of species through mutation, genetic isolation, and selection rarely leads to higher taxonomic categories, i.e., families or orders, etc.[21]

Difficulties such as these have led to postulations of a different sort by various critics of traditional selectionist views. Richard Goldschmidt has argued that evolution has not proceeded in the traditional manner as proposed by modern genetic theory. It was not the result of the steady accumulation of imperceptible mutational alterations in the gene pool; rather it must have taken place under the goad of gross chromosomal alterations.[22] These resultant "hopeful monsters," departing radically and unexpectedly from the norm, were to him the real agents of the large taxonomic movements of natural history. If we accept J. C. Willis and Goldschmidt in their argument that evolution by natural selection works downward—as we have seen there is substantial evidence for this claim—the only other solution lies in postulating evolutionary advance through essentially nonselective events. However this is still a minority opinion.[23]

Though there are still wide gaps in the explanatory structure of evolutionary theory, generally, and especially with respect to man, we do not wish to imply that the process is not amenable to an analysis within the general lines of the Darwinian model. That is, there is no reason to attribute evolution to those traditional metaphysical categories of final causes or inner-directedness toward a preestablished spiritual goal that some scientists and philosophers have proposed as a means of avoiding the mechanical and random character of evolutionary advance.[24] However, within the present structure of the theory of evolution, we can attempt to show the possibility of accounting for the creation of the human brain as a process independent of the slow and continuously molding effects of natural selection.

As we have seen, the brain increased in size from 600 cc. in the late Pliocene to 1350 cc. (homo sapiens) at the end of the Pleistocene. If this rate of increase occurred at each mutational stage by an adaptive and selective process geared to meet the survivalistic demands of the environment, it would be difficult to explain why that which had functioned as a means for the direct survival of the species for at least a million years suddenly began in the last 25 thousand to 50 thousand years to serve a new and primarily cognitive and symbolic function.

Possibly the interaction of the process of inner development, the growth of the brain, and the external forces of natural selection was in abeyance with regard to the brain's adaptive value or usefulness as contrasted with its internal cultural functions. It is important to note that the theory of evolution does not necessarily tell us about the exact sequence and tempo of the process. It merely schematizes the logical factors which enter into the process, each factor having a relatively autonomous role to play. Thus the variable of mutational change is conditioned in its effects by the adaptive capacity of the species. This in turn has been determined by the total physiological and morphological structure of the species as it has been in process of developing over many millions of years. All of these factors are finally acted upon by an environment which itself may be in the process of change. The rate at which the selective forces will act is thus a product of these variables. This situation can only lead to an infinite number of possible selective rates. In theory then, if we consider that the evolution of life has been taking place for over two billion years, for a million or so years certain genetic changes might well have not been under the strict supervision of natural selection.

Biologists have noted several ways whereby such events could have occurred. The evidence of course is somewhat theoretical, even hypothetical, for these events can be examined and imputed for the most part only indirectly from the fossil record. There are perhaps three different ways in which one can explain evolutionary changes of this sort without invoking the rigorous shaping forces of natural selection. The first is through the concept of linked genes.[25] We know that though genes mutate independently for the most part, the phenotypic expression of certain genes occurs in pairs or in even larger groupings within the chromosome. Thus it is possible for a phenotypic expression of one gene to be positively selected, and if its associated gene is not negatively acted upon, the latter will continue to develop in tandem with the actively selective trait. Thus if in one species color and shape of an organ are associated genetically and if color plays a selective role, any selective increase in intensity of color will carry with it the intensification of the phenotypic expression of the gene for shape. Yet the latter is not being acted upon selectively. It is neutral in this situation. In the history of evolution, it must be axiomatic that the vast majority of genes do not play an active role in the adaptive process at any one time.

If at some subsequent time external conditions change and the shape of the organ assumes an important adaptive role and color becomes irrelevant, then we will have had the creation of an organ whose character is very different than in the earliest stages of its evolution and yet the changes associated with this now important adaptive feature would have been by and large by-passed by the forces of natural selection. This works in a nega- tive sense also and has probably led to the extinction of many creatures. An example is the Irish elk, whose huge antlers were a product of differential growth of the body. When the elk grew in overall size the antlers increased proportionately, only to con- stitute in their annual replacement too great a drain on the phys- iological viability of the creature.[26]

A second possibility we call preadaptation. In the strict sense all changes in a species are preadaptive because mutations are random; if an alteration works out to the benefit of the creature it would have to take place subsequent to the mutational change. Those characteristics then that are usually ascribed to preadap- tation in the evolutionary process exist in some form over a longer period of time but are not acted upon by natural selection. Thus it is possible that recessive genes, not having significant selective value, could become positively "charged" and be transformed into dominant characteristics. Similarly a gene might have a negative selective valence and be suppressed in the genetic struc- ture only to be endowed with a positive value at a later date. The fact that these genes exist as a resource for the creature to draw upon at some future time allows us to label this as a preadaptation.

Again, if we follow Goldschmidt, a large mutational change might occur which would give rise to a phenotypic characteristic out of all proportion to the current adaptive dynamics of animal and environment.[27] Perhaps selection does not act rigorously on this unique characteristic at the moment. If at a future time the situation changes and selection works to the advantage of the trait then preadaptation has been established. Typically, pre-adaptation occurs when in the face of a constantly changing mosaic of geno- and phenotypic alterations, natural selection acts as an indifferent factor in facilitating or inhibiting their spread. When at a later date selection begins to give vigorous advantage to one or more characteristics, the existence of these features is considered to be a preadaptation.[28]

A third factor, traditionally the most controversial, is that of orthogenesis. In many interpretations orthogenesis—or evolution in a straight line—is conceived of as being rooted in an inner-directed trend of the biological line itself. According to this theory the process of evolution is not guided by natural selection and the play of random mutations and environmental adaptation, but rather by an intrinsic and predetermined trend of the line itself. Some apparent examples from the fossil record are the evolution of the horse, coiled oysters, the saber-toothed tiger and the early mammal *titanotheres*. These evolutionary trends, both adaptive and on occasion subsequently nonadaptive, were once thought to be unexplainable through the theory of natural selection and random mutations.[29] While most of these developments have been shown to be nonselective in the short run, their right to be incorporated into the mainstream of evolutionary theory now remains unchallenged.

One does not have to argue in a mystical or metaphysical vein, however, in order to account for a nonselective "orthogenetic" trend on the basis of recent knowledge. As we have already noted, natural selection has to act on the store of the random mutations which have occurred in all directions. True, after two billion years of genetic development there are certain possibilities for change which have been lost to all contemporary living creatures. In fact, since most mutations are deleterious, as we have noted, the rate of mutational change itself has been selected downward. Thus if any one organ or feature of a creature is subject to strong positive selection, other structural or functional mutations will consequently either be neutral or subject to elimination. After perhaps several generations, the positively selected feature, e.g., brain or body size, reproductive efficiency, length of tusks, will tend to mutate again in the same general direction and if selected positively will be reinforced once more in the genoplasm to the constant and increasing inhibition of other possible mutations. As the groove gets deeper, by the very laws of probability, one set of mutational alternatives will occur more often than others. George Gaylord Simpson has labeled this process as "ortho-selection." This would constitute one explanation of those long-term trends observed in evolution. Natural selection and adaptation are of course still operative.[30]

3

Let us suppose then that the selective factors which had been acting on the prior and long-term trend now become inoperative. Are we to suppose that the genetic trend that had been allowed to dominate heretofore would likewise automatically come to a halt? This is doubtful, since mutation rates are not subject to external influence.[31] It is reasonable to assume that, as with man's brain, the trends which were selective at one time (through the Miocene), continued during the Pliocene and Pleistocene and that while a 600 cc. brain was perfectly adequate to meet the demands of the environment, a 750 cc. or a 900 cc. brain, when these accretions developed, was equally adequate. The new morphological features were adaptive in the environment, but the lack of existing environmental restraints could have acted to allow the number of hominid varieties to proliferate immensely —as occurred with Darwin's finches on the Galápagos Islands. Thus along with a fortuitous ecological opportunity in the environment, there ought to have been a parallel increase in variations in the kinds of hominids now appearing as well as a sharp change in the tempo of evolution (towards Simpson's rapid, tachytelic level).

This characteristic of rapid advance, concomitant with the creation of radical ecological opportunities, is further exemplified in the prior proliferation and variation of the mammals at the beginning of the tertiary period, which followed upon the extinction of the dinosaurs.[32] Thus the crucial factor in any great evolutionary advance on the taxonomic scale is the enlarging of environmental opportunities far beyond prior selective limitations of the earlier era, for any one or a number of groups of living creatures.[33]

There is yet another line of research, that of Sir Gavin deBeer, which has opened up some very illuminating conceptual possibilities, buttressed in addition by a wealth of empirical support. It represents perhaps the most illuminating synthesis of the diverse strands of research dealing with the embryological and ontogenetic aspects of the evolutionary process.[34] DeBeer has acquainted us with the effect of heterochronic mutations—mutations which affect the development rates of organisms. In effect his research has opened up the possibility of answering these two questions: (a) how homo sapiens could have developed in such a short space of evolutionary time, and (b) how movement up the taxonomic scale can be explained through a genetic theory of morphological change.

The discovery of these genes which effect changes in the morphological structure of the organism by speeding up and slowing down the process of development without essentially adding or subtracting organic structures has led to the development of certain concepts of great heuristic value. In general, terms such as *neotony, paedomorphism,* or *foetalization* all refer to a process whereby characteristics in a modern adult form remain in the same general condition as the foetus or young of the ancestor. Or, for example, there are animals which become sexually mature without essentially changing their larval form or habits.[35]

The application of these ideas to the study of human evolution was first undertaken by Bolk, who noticed the amazing correspondence of the modern adult man with the young of the various anthropoid apes.[36] A large number of uniquely sapient features such as erect posture, shape of feet and toes, lack of body hair, absence of heavy brow ridges, undeveloped jaws and thus childish face, pinkish skin (in Caucasoids), and finally the lengthy period of immaturity together with the delay in the closing of the sutures of the skull, seemingly related to the expansion of the new brain (frontal lobes, cortex), are characteristic of the anthropoid young. All the evidence indicated that these factors were effected through a massive readjustment in the time schedule of ontogenetic development in the ancestors of homo sapiens.[37] As deBeer states:

> It has to be concluded therefore that in the evolution of man from his ancestors, neotony has taken place in respect of several features which show foetalization. At the same time, of course, in other directions, the evolution of man has involved progressive change of vast importance, some of which, however, might not have been possible (e.g., the development of the brain) had it not been for certain features of neotony (e.g., the delay in the closing of the sutures of the skull). This case has been described fairly fully to serve as an example of the phylogenetic effect which heterochrony can produce by allowing characters which had been juvenile in the ancestor to become adult in the descendant.[38]

The overall tempo of hominid ascent as well as the existence of gaps in the fossil record make it difficult to conceive of this process of development in the evolution of man as being other than what Goldschmidt called the creation of a "hopeful monster." The fact that this process of paedomorphism puts a premium on the selective value of immature characteristics, those protected from the environment, also gives support to the possibilities of a trans-selective process. Thus where new characteristics first appear in the young and then undergo heterochronic alteration, they may at first be freed from a selective sorting to find their way into the adult forms in rapid succession:

> Neotony following on the evolution of caenogenetic characters may result in the sudden appearance of new characters in the phylogenetic series of adults; a condition described as "clandestine" evolution. It follows that paedomorphosis will also be associated with this type of phenomenon; and as the characters undergoing clandestine evolution will not appear in the phylogeny of adults for some time, not only will such characters make their appearance in phylogeny suddenly, but, before they do so, the fossil record will show a gap.
>
> It is to be expected, therefore, that in those cases where paedomorphosis had occurred by means of neotony, resulting in the production of markedly new types, such as the chordates, much of this evolution will have been clandestine, and there will be a gap in the record of fossil ancestors of the new type. . . . In other words, the theory of paedomorphosis not only explains the gaps in the fossil record, but also supplies the reason why such gaps must be expected.[39]

The above material allows us to venture several tentative suppositions concerning the development of man's symbol-mongering brain. It is thought that ecological upheavals of the late Miocene or early Pliocene brought about the deforestation of large areas of the world and the creation of great treeless plains. This, together with the selective threshold which man's hominoid ancestors had crossed, might have set off a frenzy of evolutionary activity. New opportunities were opened up with the genoplasm presenting a repertoire of novelties. In this development of "genetic instability," the tendency toward heterochrony and foetalization, which had already manifested itself in earlier genera of the order, ran rampantly ahead, especially since in this new environment selective requirements were as yet ill-defined within the now broadened ecological niche of this group. It should be noted that the popularly conceived forms of natural selection, e.g., interspecific competition of the tooth and claw variety, have always been overexaggerated. It is not open warfare between species that usually occurs within a given ecological niche, but a more subtle form of competition for the life space. Differential reproduction is the key to victory or defeat for a species.

The absence of any clear-cut ancestors of homo sapiens in the fossil record can be explained in two ways, neither mutually exclusive. The first would echo deBeer's contention that the process is sudden and is covert in the evolutionary process—in that changes in the foetal or youthful stages are difficult to ascertain —until they become part of the adult morphological structure. Then we can expect that the development of the foetalized species would occur suddenly, as suddenly as the appearance of homo sapiens about 35 thousand years ago. This might also explain why man has remained relatively stable genetically in the ensuing period, i.e., he came forth as a sudden, hopeful, and fully formed "monster." The neotenization had completed itself.

A second possible explanation for man's sudden appearance can incorporate the first explanation but pushes the process backward in time. Man need not have sprung out of the mists of the late Pleistocene in a delayed burst of brain development, as Eisley and others seem to suggest, but may have been suspended in a selective limbo for many millenia while learning to cope with his newly acquired symbolic capacities. This latter hypothesis is predicated on the fossil evidence that may place a precursor of man—Swanscombe and Fontéchevade—in an earlier part of the Pleistocene.[40].

4

The change to sapient status was accompanied by a vast ballooning of the new brain, the cerebral cortex and the frontal lobes. As a consequence the older instinctive forms of automatic behavior were inhibited. Now with the domination of the "silent" brain, the new and integrating system of human responses was shifted from the signal to the symbolic realm of thought and action. What had begun as a neural loop attached to the olfactory lobes in the late Mesozoic as a means of slowing down or intercepting sensory messages to the system of afferent and efferent nerves—the reflexes—in which the circle of sensory input is immediately transformed into physical or motor output, had in the late Pleistocene acquired a completely different function.[41] The new brain not only had vast powers to delay and integrate an infinite variety of perceptions over an extended range of time, but it was sensitive to a greatly increased spectrum of stimulations. As Susanne Langer has put it:

> In man, nervous sensitivity is so high that to respond with a muscular act to every stimulus of which he takes cognizance would keep him in a perpetual St. Vitus's dance. A great many acts, started in his brain by his constant discriminative perception of sights, sounds, proprioceptive reports, and so on, have no overt phase at all, but are finished in the brain; their conclusion is the formation of an image, the activation of other cell assemblies that run through their own repertoire of word formation or what not, perhaps the whole elaborate process that constitutes an act of ideation. . . . The completion of peripherally or centrally started acts within the brain usually strikes into other cerebral events; and it may have been an intolerable crowding of impulses that finally led to the most momentous evolutionary step in our phylogenetic past, the rise of spontaneous symbolic identification of percepts, recollections, and free images, or figments with each other, which grew into a characteristic and pervasive tendency.[42]

The freeing of man's ancestors for a short and momentous period in the history of life from the rigorous and closely guarded selective demands of the environment and at a crucial moment in their own evolutionary development thus allowed that great and unique morphological accident of history, the sapient brain, to come into being. Its creation has brought forth a wholly new set of categories in organic existence, as had previously occurred with the earlier creation of chlorophyll, the spinal chord, and the amniotic egg. Factors such as language, man's "time binding" capacities, his ability to objectify both inner and outer forms of sensations (body and mind) and the cultural correlate of this generalized capacity—art, science, and philosophy—are not manifestations of man's break with biological nature, for in fact these are all products of his unique biological heritage.

This radical biological tradition is a product of a crucial morphological change which occurred as one of the more unique, if not gross, accidents in the history of life. What had presumably worked so well in securing the well-being of the order in the forty or fifty millions of years since the earliest tree shrews climbed up off the ground, i.e., the development of cortical intervention and evaluation of sensory messages, had by the sudden expansion of these cortical accretions sometime during the Pliocene or Pleistocene created a new level of biological functioning.

As a result of this radical structural alteration, no traditional instinctual or survivalistic animalian functions, whether physical combat, sex, nutrition, even the preservation of life itself, could ever again be independent of cortical control or involvement. We have a conditon, as Cassirer noted, wherein all human responses are lifted onto a new adaptive level. All man's behavior now becomes subject to the evaluations and considerations of the new brain.

The growth of the cortex, even as a relatively sudden evolutionary phenomenon, certainly took place over a large number of generations. It was during this period that man must have learned, perhaps painfully, to shift his behavior from an instinctual to a societal locus of control. The morphological potentialities of the brain are certainly responsible for translating man's inner thoughts into an external and public form of behavior—language. Man's convenient capacity to vocalize and thus express symbolically his unique rendering of perceptual experience through meanings communicable to others facilitated the growth of cultural control. Thus the first phase of an adaptive shift was culminated. The second, by which the symbolic realm itself began to be utilized for man's survival and adaptation, must be thought of as the subsequent use of a preadaptation, the primary purpose of which was other than a concrete organic response to an external environmental challenge.[43]

We have noted earlier in this study, in regard to the problem of aphasia, that what Kurt Goldstein called the abstractive qualities of ordinary normal thought and action, is centered on the very structure of the human brain. Injury to the auditory centers of the brain results in deafness for man. He, unlike some lower forms of life, is unable to shift his ability to hear to his brain stem.[44] Man's unique powers are completely concentrated on this new level of symbolic and abstractive thought. An individual with minor brain injuries can still function automatically, e.g., he can apply a fork to food set before him or turn a doorknob that his hand has been placed upon, without ever understanding the nature of the objects he is handling, but he is no more a man. Physical acts as such, even when they lead to biologically functional behavior (the concrete attitude), are not included in the circle of human cultural behavior. Here the pervasive theme is meaning and structure, i.e., the integration of experience through conventionally agreed upon symbols (Goldstein's abstract attitude).

The question can be raised as to the extent to which the Darwinian constructs of natural selection and biological adaptation are contradicted in this postulation concerning the development of the sapient brain and the creation of cultural behavior. The answer will depend upon whether this theoretical maxim is held to apply to every moment in the historical process which it purports to explain. Since it is a historical theory, the element of time plays quite a different role than it would in physical theories of a structural type.

Because historical time is a smoothly flowing process that can be subdivided almost indefinitely, physical theorists seek to express their formulae in such terms as to neutralize it. In physical theories time flows back and forth across the equation as if it proceeds in any or in no direction. Historical theories have to account for the unidirectionality of time. For most historical laws the prediction will state that certain events will occur over a given span of time. It is virtually impossible to set forth every historical law so that the predicted events will be seen as taking place at all moments in this temporal continuum. No experimental canon of confirmation could validate such a claim. Therefore, it is logically admissible to formulate such historical theories and yet allow for certain temporary counter causal processes.

Merely to postulate that an occurrence will take place in a given duration allows a measure of indeterminacy with regard to events within the duration. So long as the requirement that "x" will occur at the termination of this temporal rhythm is fulfilled, we can tolerate momentary counter causal or nonpredicted events within the theory. To exclude the possibility of this happening $(t=0)$ in effect excludes the existence of historical theories of this sort.

An interesting and important example of this process in the physical realm is given by the second law of thermodynamics, first enunciated by Sadi Carnot in the early nineteenth century. Labeled by Sir Arthur Eddington as "time's arrow" because it points in a specific temporal and therefore historical direction it is thus a unique species in physics. The second law thus occupies a special niche in several disciplines. The classic formulation of the law by Clausius (1854) states that in any closed physical system, the energy of the system in time will always show an increased degree of randomness or lack of orderliness.[45] This can be characterized as reflecting a steady diminution in the concentration of energy and the consequent even dispersion of energy throughout the extent of the system. All thermodynamic systems in time show an increase in entropy (their level of disorderliness).

But living processes exemplify an opposite tendency. They obtain energy from varying sources, organic and inorganic. In feeding, growing, and reproducing, living creatures exemplify a counter-entropic process. For the most part the energy sources given to the planet Earth at its inception are gradually being dissipated into the environment. At some future date this exchange of energy will be concluded. The processes predicted by the second law will have been fulfilled.

While the total energic degradation of this one area of our universe will have proceeded a local counter-entropic process will have occurred. As long as living things endure they will build up their energic potential from the available resources in the same basic manner as the earliest eo-bients took their nourishment from the protoorganic sustenance around them and thus prevented their own dissolution.[46]

Does this, however, entitle us to say that the second law of thermodynamics is invalid in this instance? Certainly not. First of all we must designate our total energic system. If we stipulate it to be the solar system we would probably find that more energy is being diffused into the region of the system by its various centers of energy than is being concentrated in any one portion. Thus when one considers the entire thermodynamic system, the drift toward entropy is corroborated. The law does not specify that this process must occur uniformly and constantly at every point and moment in the entire energic system. Thus it is theoretically possible and indeed is existentially true that temporarily counter-entropic processes can take place without contradicting this principle. Certainly throughout the entire system a greater amount of disorderliness is being created over a constant flow of time. And more energy is being dispersed into space than is being utilized and thus structurally trapped for further use by living organisms.

We can explain in a similar manner the fortuitous and remarkable growth of the human brain and still not fault the theory of natural selection. The specific structural characteristics of the brain are only generally the product of close external selective supervision. Its existence at the end of the Pleistocene was only partially due to its adaptive utility. The Darwinian theory is certainly applicable in that at the end of a long temporal period the hominid brain was an adaptive success. It existed, and thus fulfilled the gross predictive requirements of the theory. The theory never asked how long this process of adaptation and selection would take nor the intervening factors and events between the two points in time.

A further complication is added when we note that the morphology and physiology of the sapient brain tell us little of its potential for biological adaptation and its utility in the survival of the species. As an expansion of the lemuroid brain, it might be conceived as an "escape facilitator" par excellence. But now the vast neurological accretions, instead of processing information for evasive physical behavior, produce language and culture. On the face of it, language and culture have no clear biological meaning in terms of the survival of the species. Evolution gives us no parallels for understanding the symbol system in terms of its physical efficacy in nature.

It is at this point that many errors have occurred in the interpretation of culture. Obviously, the universality of culture must reflect positively on its adaptive utility for man. Though we might say that without culture man could not survive—bereft as he is of specific instincts to meet his gross biological needs—culture itself is no sure guarantee of survival. Further, the multitudinous artifacts and attitudes which constitute culture can be as destructive to man as they can be productive. Symbolic behavior has been presented to man as some ominous, opaque gift whose ultimate nature will be determined by man's own use. No other instinctual or adaptive skill has operated as such a two-edged sword.

Tribes and civilizations have literally disappeared because of their inadequacy in adjusting their conventional meanings to changing circumstances. Also, we must add that ubiquitous and unexplained fact that men kill each other, destroy societies and civilizations as much and as often as they cooperate to secure the best interests of their fellows. There is thus no necessary correlation between the existence of culture and any clear-cut and specific adaptive and survivalistic function that it may embody.

The theory of evolution implies that nature does not create organs or entities that are purposively preadapted to the environment. Nature produced the eye, the wing, or the lung, organs that almost immediately helped some creature to secure itself in our world. But it also has produced countless other novelties for which no use was found and which eventually disappeared. The singularity of the human brain lies in the fact that it was allowed to develop to a point that it now articulates and extracts from human experience a totally new set of natural characteristics, those associated with culture. These characteristics, as they are manifested for example in the diverse forms of knowledge, now refer to biological needs of a wholly new sort, i.e., the problem of meaning, which the traditional vocabulary of evolutionary theory is as yet incapable of describing and subsuming.

Is there then any evolutionary significance to the constantly changing panorama of scientific theories as it shifts from invariant to invariant, always seeming to break down our pretensions to an objective and permanent world of truth existing either in the content or the form of our theories? How can we explain that which seems to remain permanent and firm as a methodological demand of thought on experience, namely, the quest of man to universalize his experience by achieving wider ranging theories (the search for the Parmenidean Being)? How do we explain man's dissatisfaction with the imagistic, material, and ultimately particulate symbol, in favor of greater abstractedness, his need to substitute thought for image?

These questions cannot be definitively answered of course. We are confronted here with a problem of the structure of thought, itself a product of the same nature it seeks to explain. Obviously the cognitive demands man makes of nature are due in some manner to the new structures of his brain. It is probably warrantable to assert that in the not too distant future we shall have a greater knowledge of the neurological correlates of the higher levels of thought. But here we arrive at a typical Kantian problem. Suppose we do achieve a close association between the symbolic and externally representable manifestations of thought—as given in the sciences and philosophy—and the symbolic representation of the neurological pathways which are invoked when the mind addresses itself to these diverse subject matter contents. Would we be entitled to say that we have reduced our knowledge of experience to neurology? Would we not be equally entitled to argue that we have reduced neurology to epistemology? Could we ever expect that a complete knowledge of the molecular interactions of the mind would therefore limit the dimensions of novelty in the realm of thought, or vice versa?[47]

We have here the basis for Cassirer's contention that we should not seek to reduce knowledge to any externally or internally rooted substantialisms. Indeed, in the above example, the neurological analysis of thought is itself a symbolic creation of man—a set of theories and concepts. Rather than existing as a more basic ground of analysis, when examined in the context of the creature —man—who forms these categories of knowledge, it has no essential logical priority. Only man, *animal symbolicum*, has any ontological or noumenal significance. And as Cassirer has pointed out, the only way we can know man or define him is through his works. Not only can we not know the nature of these works in advance, but inevitably they are recreated and redefined at every new instant in time.

9

The Direction of Knowledge:

A Critico-Idealist Exploration

BEFORE CONCLUDING OUR STUDY LET US SUMMARIZE
the major claims of Cassirer's position:

1) When it is sheared of its debilitating *a priori* commitments,
Kantianism, in asserting the autonomy of thought and knowledge
from traditional reductionist and realistic metaphysics, is applicable to knowledge at all stages in its historic development.

2) Knowledge follows thought. Once physical science is freed
from metaphysics, certain logical motifs of theory construction
can be traced in their development. These motifs have their analogies in the growth of language and in the structure of other
cultural disciplines. The Kantian thesis that thought imposes its
canons on experience, rather than vice versa, is exemplified phenomenologically through the disciplines of knowledge rather than
through any ontological explanations.

3) The significance of man's pervasive theoretical endeavors
is set contextually in the conception of symbolic man, a position
that is opposed to traditional selectivistic and adaptational views
of human behavior. Theorizing—the search for meaning—whether
it be common-sense behavior or mathematics, is seen as an intrinsic preoccupation of the human species.

4) Through his historical approach to the problem of knowl-

edge, Cassirer was able to delineate the slow but persistent process by which thought is freed of metaphysical misinterpretations of knowledge. But he has also revealed the opposite side of this intellectual coin, illuminating an interesting recalcitrance in thought. For with each intellectual breakthrough, with each reintegration of the main themes of knowledge, we find the same substantivistic, reductionist, and realist aspirations. The persistent search for one permanent truth or one ultimate criterion for objectivity is mirrored in the realm of scientific thought by a search for similar epistemological categories. It was thought that this type of search was rejected when man emancipated himself from traditional absolutistic theology. The lesson, of course, is that we must not be bemused by alterations in the formal or exterior qualities of knowledge. While the language of metaphysical thought may change and thus superimpose itself on a modern and innocuous facade, the essence of metaphysics must still be eliminated from any realm of knowledge which may fall prey to its easy securities.

5) Perhaps more than any other contemporary philosopher, Cassirer illustrated the importance and utility of going to the empirical disciplines themselves for the evidence necessary to deal concretely with the various issues in the theory of knowledge. The scope of Cassirer's learning—both in range of information available in the disciplines and in his knowledge of their inner theoretical structure—was such that it is hard to find an area that escaped his insatiable curiosity or transcended the grasp of his great intellectual powers.

It may well be that this universality of perspective together with his continental literary approach was the cause of Cassirer's relative obscurity as compared with other twentieth-century philosophers. In an era in which the disciplines were becoming increasingly specialized and withdrawn from the more general concerns and scrutiny of philosophers and in which, on the other hand, philosophers themselves were more preoccupied with their own parochial professional issues, Cassirer's contextual and synthetic concerns may have appeared somewhat atavistic.

Cassirer was a traditionalist in philosophy in that he was concerned with the problems of an intellectual community, problems which lay for the most part outside philosophy. He subjected them to philosophical scrutiny without ignoring the contextual and experiential base in which they took place. And while his systematic philosophy, as distinguished from his historical studies, was focused on epistemology, his last book, *The Myth of the State*, represented an important application of both these strands of his career to the problems of contemporary political theory.[1]

6) Perhaps most important, Cassirer presented us with a new set of problems and questions that had not previously existed in the theory of knowledge. In this he brought us to a new philosophical threshold, not in any sense to belabor old and outworn motifs in philosophy, but to reexamine some of the perennial concerns of man from this new perspective. By shifting his epistemological point of view to include not only philosophy and physics but also culture and the problem of evolution and by focusing on the crucial questions of man's nature, behavioral motivations, and thought, Cassirer was able to give the problem of social and historical change an entirely new perspective.

Rather than examining culture from the standpoint of its materiality or its intractable realness, which demands subjugation and conquest in the name of man's adaptive and selective needs, rather than viewing the forces of technological and economic change in terms of an external reality existent and irresistible in terms of human choice, we can now see all of culture, its structure as well as its dynamics, as an instrumental creation of thought. If technology seems irresistible, it is not because of its lack of amenability to human intervention, or because of its supposed reflection of an ontological order of things. Rather, technology represents an aspect of man's symbolic drive. To understand this force in terms of human cognitive needs would perhaps enable us to place it under the control of the human will. This can never happen as long as we conceive that the creation of culture is rooted in external and permanent metaphysical principles.

But this is only a suggestion of a problem and an approach

which Cassirer never fully articulated or developed.[2] The richness
and potentiality of viewing man as a symbolic creature are so
far-reaching that we cannot hope to explore all the possible avenues of philosophical investigation.[3] We will conclude this study,
however, with a brief examination of several fertile but as yet
undeveloped and unexplored issues.

1

CULTURE

The most intriguing question that is raised in merely postulating man's symbolic capacity is how culture and knowledge
came to be. How, from an inner propensity to express thought
symbolically, do we get the diverse forms of knowledge and culture manifested by history?

There are many ways to approach this question. Our purpose
is to examine how Cassirer approached it. As we have seen, Cassirer's tracing of the phenomenon of symbolization could be characterized as working downward from the upper branches of the
tree of knowledge. Merely by examining phenomenologically
the structural characteristics of each of the symbolic forms, from
philosophy and physical science to the various cultural and informal disciplines, we cannot realize that the crucial problem is the
overall structure and meaning of this system of thought.[4]

Not until he wrote *An Essay on Man* did Cassirer put aside any
possible imputations that the divisions of thought he developed
so naturally in his prior and more systematic writings were to be
thought of in any way as fixed substantial realities of thought or
culture. Thus in *The Philosophy of Symbolic Forms* one can see
Cassirer arriving at a position in which language is placed in the
central position in the evolving spectrum of symbolic development. From the capacity to speak come the possibilities for the
unique tri-fold division of thought which Cassirer traced in its
structural evolution. Basic to man's use of language are his secu-

lar common-sense usages which, with unfolding self-consciousness, eventually develop into philosophy and science. In addition the vibrant and emotive images of proto-mythic and religious symbols are first given expression through the magical powers of words. Finally, the poetry which lies potentially in all of us as a capacity for esthetic envisionment cannot be objectified until it finds its wings through the word.

But this division is not strictly held to in Cassirer's later writings. At times language is placed in its own special category as distinct from the more formally discursive forms of thought. At other times it is given the more orthodox interpretation in which its general genetic or historical affiliation is raised. In a third interpretation common-sense language is united genetically to the exact sciences merely as an undeveloped form of the latter.

As we have noted, Cassirer abjures even the most remote possibility of substantivism when he adds to his schematism history and biology as distinct forms of thought.[5] Still, if we look deeper we find the synthetic element which makes possible the entire cognitive enterprise. Arching over the entire symbolic edifice in this philosophical anthropology is, as Bernard Groethuysen has noted, man's capacity to philosophize—the ability to appreciate and order logically his own heterogeneity of thought.[6]

Throughout Cassirer's writings one notes the search for an ordering principle which would characterize each symbolic form. In the discursive realm, for example, every perception and vocal unity is immediately subsumed under and integrated into a larger, more universal structure of meanings. The desire for order, for logical and causal relations, dominates every area of experience which will eventually come under proto- and fully scientific modes of thought. In myth and religion the unitary principle is derived from feeling or emotions. Here secular or logical principles fade in effectiveness. Expressive and affective societal, cosmological, or personal images seem to exhaust any further external reference in adding to the significance of the symbol. But even in this domain, as mankind develops more sophisticated societal forms, religion and myth are invested with greater ideational and abstractive contents. On the level of monotheistic religions, the emotional categories which undergird even the earliest phase of religious experience are never absorbed into science, eventually to be lost to religious thought. They exist on a different plane, separate from but equal to man's logical world of cause and effect. In this way they carve out for man a permanent role in his psychic and cultural life.

Art also emancipates itself from basic mythic sources, according to Cassirer. As man matures culturally he recognizes implicitly that among his intellectual and symbolic needs there resides another realm of expression with its own unique intentional vector. Cassirer describes art as satisfying man's sensuous involvement with experience. The postulation is made that the existence of each sensory organ provides an opportunity to achieve a measure of objectification. The possibility of an esthetic realm is therefore created. Each of the symbolic modes expresses its own special logical principle. Each has an inner life of its own.

Later Cassirer added other symbolic forms to his structure. He saw that in all societies, man's universal sensitivity to the temporal dimension of experience had resulted in the creation of the idea of history. The ability to recall to consciousness either the personal or social events of the past and to integrate disparate fragments of time into an articulated interpretive whole constituted a real and recognizable aspect of human knowledge. Whether formulated in terms of mythic and religious symbols or in the modern discursive representations that perhaps began with Herodotus, the fluid interactions of past, present, and future on the minds of men, so removed from the practical temporal exigencies of lower organic forms, reflect one of the most important structural dimensions of human thought.

Biology also began to occupy Cassirer's attention. As he examined the systematic principles of this discipline he found that here too a new set of ideas had created a new dimension of knowledge. Principles of organic structure (morphology) and function (physiology), taxonomic development and evolution, in spite of their indubitable roots in the physico-chemical structure of matter, had to be dealt with and understood in terms of their special conceptual character.

In general, Cassirer's *Essay on Man* is not an attempt to explain the many in terms of one unifying principle, though Cassirer does emphasize man's symbolic character. However, this noting of unity in diversity is not supported with a postulation as to how the plurality of forms was derived.

One must assume that historically, as well as genetically and morphologically, the capacity for these symbolic modes of thought existed prior to their actual envisagement in human behavior. Man did not burst forth with these cultural modes wholly articulated. That the differentiation and pluralization of thought did come about is a fact. What needs explanation is the logical process by which a generalized genetic potentiality expressed in the symbolic nature of human behavior produces a cultural structure so variegated yet not chaotic or disorderly.

We are aided in our quest to explicate this cultural theory by a recent and important book of Susanne Langer. *Feeling and Form* is devoted to a general theory of art developed out of the broad survey of symbolism initiated earlier in her *Philosophy in a New Key*.[7] In its detailed manner of analysis of the characteristic motifs which delineate the varied art forms it could be considered to constitute the fourth volume of Cassirer's *Philosophy of Symbolic Forms*, a worthy fulfillment of the program that he was not able to complete.

In this book Mrs. Langer develops a position—roughly the same as Cassirer's—that the arts communicate in a genuinely symbolic but not ideational or discursive manner.[8] Her theory of the art symbol as an expressive form is developed here in great detail. Through it she emphasizes both the individual artistic disciplines and a wide-ranging segment of the critical literature in the arts.

She does not go beyond Cassirer when she establishes feeling as the unifying theme or category which distinguishes presentational symbolism from the discursive realm. As we have noted, Cassirer made a tenuous distinction between mythic symbols, which encompass feelings, and art symbols, which are unified in their sensuous functions. But since Mrs. Langer makes a more inclusive categorization of symbolic forms, there being two general kinds—discursive and nondiscursive, the latter including both mythic-religious and artistic symbols—her differences with Cassirer are not crucial.

Thus, though rooted in feeling, art is not characterized by the expression of inchoate and disorderly emotions. Though its psychic sources diverge from the discursive and logical realms of scientific order and generality, it is immersed in its own forms and structures, with its own special logical discourse. The potency of art in culture lies in human nature itself. And while one can, in a Kantian manner, note that feeling is what is constituted by the self, to follow through on the psycho-biological sources of this emotive valence lies beyond the scope of her study. It is enough for her to note that this is not the mere biological expression of emotion, for art is always connected to an image. This image is provided perceptually. Thus the union of feeling and perceptual image constitutes the polar elements of the presentational symbol and provides the genesis of artistic import.

The variety and heterogeneity of esthetic experience derive from the diversity of the media. These provide the images that are necessary for the process of symbolization. Man's esthetic intuition flows out uniformly in all directions. It is diversified into architecture, music, painting, literature, and the dance, depending on the environmental and experiential materials available. Certain art forms demand the existence of materials like stone or marble, wood, canvas, and pigments. Some art forms express a particular encompassing of experience in spatial or tactile terms, others in temporal, rhythmic, or aural forms. The dance on the other hand involves the entire body in a spatio-temporal and sensuous relationship that creates for itself a virtually organic four-dimensional environment.

As Langer points out again and again, art does not copy experience. Rather it transforms those varied possibilities inherent in human perception through the form-giving powers of the mind and through the emotive resources of our nervous system. The result is a wholly recognizable and autonomous structure of meaning. But, in addition, we can note that the joy and richness we feel in the thrall of the artistic experience, once we have educated our perceptions to "hear" Beethoven or to "see" Renoir, are enhanced by the other cognitive powers of the mind. Thus as we gain an understanding of the formal character of each art—as artists have learned their craft so as to create—and thus attain a measure of intellectual mastery over the various structural dimensions involved in each domain, our esthetic perceptions become sensitized to and further excited by the esthetic understanding.

The more we know of the harmonic, contrapuntal, and orchestral resources of the composer, his use of the great architectural forms of music, sonata, concerto, opera, symphony, the more we are aware of the interpretive resources of the performer and the conductor, the more we appreciate the varying stylistic nuances of such things as national style, orchestral quality, the characteristics of the instruments being used, etc., the more powerful our feelings of enjoyment, pleasure, and esthetic satisfaction will be. Thus it is never a raw physical exhilaration that gratifies us esthetically. Rather it is the accumulation of varying levels of meanings that heightens the entire experience. The core of the esthetic experience, however, is derived from an inherent and basic capacity of our senses to form perceptions (structures) and of our emotions to infuse these structures with meanings—in Langer's terms, to create "expressive forms."

It is then easy to see that in every area of human perceptual experience possibilities will exist for the creation of these expressive forms. Certainly, a few sensory areas will be more amenable to the kind of logical supplementation that we find in music, painting, poetry, sculpture, and architecture. These arts will go further toward objectification and structural development. Areas of sensory import such as the olfactory and the gustatory are also rich sources for esthetic pleasure and appreciation. But because they are more removed from our general cortical integrative powers and have a smaller discriminatory range we can say less about them. We can sniff and exclaim with pleasure, or taste and murmur our praises, but the range of logical communication is smaller. Pleasure here is more private, and the possibilities for structural development as compared with the other arts more difficult.

What remains uniform amidst this differentiation as well as mixing of art forms is a basic attitude of mind which we call the esthetic. There are many aspects of experience which are not characteristically artistic, e.g., food, mortar, sound. These can all be viewed under the aegis of differing logics, for example, the discursive, in which the categories for these similar perceptual qualities might be subsumed under nutrition, engineering, and acoustics. But that mysterious and undefined capacity of man to hear and feel a musical phrase or perceive and envision a two-dimensional spatial image on canvas is recognized and universally acknowledged as fulfilling an unparalleled role in the spectrum of man's cultural meanings.

If we accept as meaningful the idea that there is a unity in the esthetic experience which represents an inner capacity in man, then we must accept the fact that the "artistic" interpenetrates a virtually unlimited range of phenomena. We would suppose that, as technology develops, new media of expression will evolve, just as photography, the cinema, television, and radio afford new opportunities for that same basic proclivity of thought—the esthetic experience. Similarly we would predict the gradual change or demise in the character and function of older art forms such as painting.

The kinds of differentiation which we experience in the esthetic domain are paralleled in the other symbolic forms. Certainly when we examine the varying dimensions of religious thought or of science we can note how the same basic logic finds its varying realizations depending upon the area of perceptual experience toward which we direct our thought. We can ask what is the significance of this process of diversification into symbolic forms and thence into varying perceptual and experiential areas of expression. Then again, we may be speaking here of something which cannot be analyzed outside further reference to the self.

For example, suppose we invoke the work of those neurologists who are developing the discipline of electroencephalography. It has been noted that it is possible to represent our inner emotions electrically and graphically. In this mechanical way differences can be distinguished in personality type, differences which we intuit but which we are rarely bold enough to state explicitly.[9]

Such factors in human experience were heretofore thought to be part of the inexpressible. Could we say that to have our inner feelings of emotion and satisfaction on listening to a Beethoven sonata translated into graphs produced by an electroencephalograph is more real or objective than the inner qualities we have felt? Suppose we were able to translate these feelings into a set of biochemical equations which would describe the chemical changes of our nervous system at the instant of esthetic appreciation. It is doubtful that we would ever have a basis for stipulating hierarchies of objectivity in these varying kinds of symbolic and expressive moments. They each serve a cognitive purpose, but the purpose is essentially incommensurable. What we have achieved is a richer way of "looking" at the same symbolic moment.

Let us add another dimension to the problem of rendering this schematism more intelligible. We have noted two structural divisions: (1) the psychological mode of creating diverse symbolic forms and (2) the perceptual differentiations which occur within each form. Our example of the latter process was given in the discussion of the esthetic. Let us add to Cassirer's original tripartite division—science, religion, and art—possibilities such as the biological, historical, social, and economic. Then let us place them into the dynamic contextual interactions in life, where their particular psychological character may become blurred and sometimes even effaced.

Now, if we are committed to a truly symbolic conception of human nature and culture, we can stipulate that this dynamic interaction is a product not of material or exterior causes but of the demands placed by reason on the evolving character of each of these consequent cultural forms. To understand the history of man, we must understand the dynamics of these varying symbolic forms and their contextual intertwinings in the fabric of culture. Further, when we note that these differing logical strands are also altered and influenced by technological innovation, military adventurism, and the as yet undeciphered symbolics of dreams and other preconscious elements, we get a hint of the complexities involved in attempting to subsume this process under cognitive control. Yet if history were intrinsically so diffuse would we still be able, after so many thousands of years, to identify or trace the growth of an idea as abstract as symbolic thought as it is reflected in each of the major cultural categories? We can, within limits know ourselves. This more general knowledge, philosophy, is predicated on the same autonomous power in human thought which forces its way to the surface of culture, whatever the context, and establishes its diverse institutional frameworks.

If we can gain a better understanding of this internal cultural dynamic as it has taken place in history and has catapulted certain cultures to great florescences of achievement, others to aggressive and totalitarian ends, and caused still others to stagnate and fade from the stage of human events, we will begin to approach the kind of inner cognitive control necessary to realize our own historic intentions.

We can understand the process of cultural development from still another point of view. Man and the world as it exists around him are products of the same evolutionary process. Man's capacity to perceive the world and the existence of a world to be perceived seem to be reciprocal characteristics on a similar logical stratum of experience. It is not surprising then that what man senses—the given of experience—and how man conceives—the structure with which he endows these sensations—merge in experience. The symbolic acts of perception and cognition thus constitute a biological function denoting a logical and experiential relationship with the environment. Their significance lies in the unique quality of the relationship, the fact that man adds a number of new cognitive dimensions to what heretofore had been a rather simple and clear-cut subject-object biological interaction. The acts of human perception and cognition are not directed toward drive reduction, stasis, equilibrium, or biological adaptation, as these have been physically conceived for lower forms of life. Man's drive is encompassed in the search for meaning, which he accomplishes in his symbolic attitude. In some way and at some time in the past the morphological accretions of the brain have been responsible for this state of affairs.

There have been attempts to explain the new symbolic functions which issue from this restructuring of the hominid brain in terms of a localized physiological activity in some particular portion of the brain. We can understand how intellectually tempting it might be to fix our cultural and philosophical explanations of the processes of mind in the simple diagrammatic neurological tracings of the electroencephalograph. However, it only obscures the problem of understanding symbolic thought in culture to contend that these kinds of explanation are epistemologically more fundamental. They merely add a new schematism which is useful chiefly in achieving a related perspective of the same experiences. Every complementary mode of symbolizing the contextual events of our day helps to add a new dimension of awareness concerning those overt and ongoing individual and social processes over which we hope to gain a measure of cognitive control.

The diverse meanings we find in culture are a product of a billion years of sensory development. Man now adds a significant new biological factor: his symbolic proclivities, which seem to be constructed of diverse structural elements with differing goals and dynamics. We observe these in culture, art, and science, etc.; we may explain them further as logical or feeling-centered elements in human thought; or we may parallel these explanations with physiological, neurological, or morphological patterns of explanation. Again our understanding of all these elements can never rise beyond a symbolic and theoretical rendering in the circle of knowledge.

Man is a predator, but on a new level of behavior and with new objects to be devoured. Every possible sensory experience that nature allows him to absorb, either unaided or with mechanical assistance, becomes grist for these diverse cognitive appetites. The need to unite, to form structures, laws, patterns of experience in every conceivable sensory mode—whether it be in the creation of a sonata, a system of geometry, a shovel and pick, or indeed a god or a senator—testify to this insatiable inner need.

2

There is a second issue, in addition to the problem of human nature, which derives from Cassirer's theory of symbolic forms and which is left incomplete and unanswered. It involves the evaluation of the various levels of civilizational achievement. Indeed, it is a problem that has concerned philosophers of history and culture from time immemorial. Cassirer hypothesized that even in man's earliest days, before he had discovered the full range of his potentialities for symbolic thought, there existed a primordial and organic unity of symbolic expression. Cassirer subsumed this under the concept of the "primitive mind." In this stage of development, esthetic, religious, linguistic, and conceptual forms of expression were not as yet seen as distinct modes of thought.

Just as linguists now believe the unit of language expression to have originally been the one word sentence, out of which the various parts of speech were precipitated, so too, on the most primitive levels of mythic thought, symbol, meaning, and object were all merged into a living and vital, indeed an exhilarating sense of presence. Experience was rife with symbols that were too real, too surcharged with energy to be given a meaning other than a living, acting, powerful sense of potency. There was no art in the sense of esthetic objectification with laws and traditions of development. Common-sense thought was as yet undeveloped and inchoate, and of course theory, abstraction, and cognition were achievements that existed only in possibility.[10]

From this state, as from the state of nebulous incandescence out of which came the solar system, modern man in all his cultural variegation ultimately developed. Cassirer was presumably aware that this earliest state of man was itself an abstraction, since all primitive societies have developed to a point where the varying spheres of life have coalesced. In all those primitive societies that we have been able to subject to investigation these diverse institutional spheres reflect an intellectual development which had presumably progressed from its ultimate primitive level. The secular realm existed for the satisfaction of basic needs. The mythico-religious realm consisted in important ritualistic and sacramental occasions—perhaps in preparation for or subsequent to wars or other ceremonial events. The esthetic dimension was manifested in dance and other choric activities, in the decorative qualities of their visual arts and crafts, and in their oral and bardic traditions.

Finally, there may have been occasions when man, in order to face new challenges, rose from the commonplace to transcend tradition. Such challenges may have necessitated decisions that taxed habit and demanded the highest of a society's wisdom. Can we doubt that even the most primitive groups have at one time or another risen beyond their ordinary attitudes and behaviors to engage, even if in a rudimentary manner, in the same discriminatory and logical canons of thought that we today utilize with respect to decision-making in science, government, or social policy?

We can accept without difficulty Cassirer's analysis that the earliest cultural levels found in our era are in themselves manifestations of a significant intellectual achievement, albeit the product of thought and social action that occurred below the surface of conscious awareness. They were reached at a time when symbolic thought was still an essentially unrealized capacity of the species. It is in our comparisons of primitive cultures with modern culture, when we postulate values of progress versus retrogression in terms of historic change, that we find real questions that need resolution.

The most obvious concerns the applicability of the evolutionary model, as it is here being broadly interpreted, to an understanding of the historical changes in culture. Is there any evolutionary significance in the civilizational advances of man, in his creations of philosophy, science, specialized art and literature, technology, and the concomitant economic surplus which Western societies have attained?

The functionalist school of anthropology, represented by Malinowski, Boas, Benedict, and Mead, in general denigrated such hierarchical or evolutionary interpretations of culture. These anthropologists adhered to a descriptive analysis of the internal cohesiveness and stability of culture and the process by which the various strands of social life were woven into the cultural fabric. Implicit in this view was the realization that all cultures that have attained a measure of stability—in spite of the variables of politics, sex, technology, and art—fulfill the basic human needs of their members. They felt that, in general, value comparisons across cultures were fraught with danger and that any particular element within culture ought to be examined first in terms of its function in each society.[11] In fact, using criteria of cultural cohesiveness, by which values and ideals are consistently articulated at all levels of the maturational process and applied to all members of their societies, it can be argued that these societies more fully realized the values and aspirations of each generation, that they exhibited less of the neurosis, psychosis, war, hatred, or perversion than besets modern societies.

As we have noted earlier the anthropologists of the Michigan school, led by Leslie White, are dissatisfied with this relativistic and antievolutionist view. Their culturological approach stresses the continuity of the evolutionary process but under new theoretical auspices. The dynamics of human evolution, they argue, necessitate uniquely cultural laws. These laws must describe the accumulation of energic resources by man, for they will be used as a criterion for evaluating evolutionary progress, which gives us a basis for transcending the cultural relativism of the majority of anthropologists. Primitive societies, since they have at their disposal relatively impoverished energic resources to accomplish their tasks in the control and utilization of nature, as compared with modern societies, can thus be characterized as epitomizing earlier stages in the evolution of mankind or retrogressive and stagnant communities on their way toward elimination (natural selection) or on the other hand nearing absorption by more advanced societies.[12]

Cassirer would have probably taken a stance somewhat in between the positions of these schools. In examining culture from the standpoint of a philosopher, he would necessarily have looked at the problem from a wider intellectual perspective. Doubtless he would affirm the necessity for a contextual and value-free orientation of the functionalist anthropologists with their delimited professional concerns, but he would have emphasized that their approach to culture is inherently narrow. The common-sense but historically significant distinctions men have made between primitive and civilizational communities are not merely the result of prejudice or present-mindedness. True, the philosopher of culture inevitably brings to bear a set of valuational criteria which obliges him to discriminate between those aspects of social life which are significant for man and those which are less significant. It is far less relevant from the standpoint of critico-idealism to dwell on and compare the varieties of social satisfactions and integrations achieved in static primitive societies and those evident in highly developed cultures than it is to note their level of intellectual awareness.

And this is the crucial point. If one values people who are happy, well-adjusted sexually, economically relaxed and indolent, one need not turn to primitive man alone. But nowhere in primitive cultures do we find the intellectual awareness, the philosophical pursuit of knowledge, the self-conscious understanding and sensitivity to the symbolic possibilities inherent in individual and society. Primitive man views society as a given which he neither questions nor objectifies. He is surrounded by an implicit conceptual barrier that binds him to the status quo of his environment. Primitive society must therefore represent not the mainstream of human development but the backwaters. What is largely implicit in human nature primitive man has barely explored or tapped. If man's nature can be thought of as being exemplified in his cultural works, then the primitive has only barely participated in sapient life.

As Cassirer pointed out, man's nature is revealed not by any one cultural or intellectual achievement, for the process of civilizational advance is a long one whose goals are still inchoately perceived. The test of human existence then lies in the manner and rate with which man unravels the various symbolic strands inherent in his thought. This cannot be achieved on a moribund cultural island. In the best sense of the term human culture, as Cassirer saw it, is "the process of man's progressive self-liberation. Language, art, religion, science, are various phases in this process. In all of them man discovers and proves a new power—the power to build up a world of his own, an 'ideal' world."[13]

This position of course represents an avowedly intellectualistic vision of man. It does not purport to reduce man to any specific state of nature or knowledge but celebrates those achievements by which civilization has been made richer, as man himself has explicitly and universally acknowledged. It is no accident then that we respect those qualities of thought which are represented in the freest exercise of the creative intellect rather than those forms of human existence which still are bound to the momentary exigencies of the physical environment and to those self-circumscribing intellectual barriers with which man cuts himself off from the future.

Our traditional idealization of Greek culture is a good example. We have gone through many phases of thought about the Athenian Greeks, from outright worship to skepticism. We have been disillusioned with their inability to maintain democracy, their incapacity for national leadership, their system of slavery, and their supposed general lack of concern with those typically Anglo-Saxon virtues of hard work, thrift, and technological progress. Nevertheless, as John Dewey once noted, what we do remember about the Greeks is their soaring achievement in the arts and philosophy and their emancipation of personal individuality. The Greeks provided us with a landmark in the history of man which is as important in what it tells us about the intellectual possibilities inherent in any one period of cultural history as in what it tells us about their own special genius.

Thus the criterion of civilizational advance—the test by which we recognize a people's exploitation of its latent creative possibilities—is productiveness, the productiveness not of material goods but of talent and genius. It is this that brings to our contemporary vision new and unexpected dimensions of symbolic meaning. Some would call this an elitist conception of culture. However, an idealistic, intellectualistic position on the significance of culture for man and man in culture is not necessarily based on an assumption that great cultural progress is dependent upon the exploitation of the masses in society. On the contrary, there is a real question as to whether great culture can exist under such conditions of exploitation—witness Hitler Germany or Stalinist Russia.

On the other hand, cultures which are sterile intellectually and esthetically are perhaps more repressive than our current canons of theoretical understanding would lead us to believe. We cannot believe that those great achievements in the arts, science, and philosophy which have come in clusters both in terms of time and location are merely expressive of a random explosion of genetic potential. In the position here being articulated the essence of homo sapiens is expressed in the work of an Einstein, a Bohr, a Darwin, or a Mendel, a Monteverdi or Beethoven, a Shakespeare, Rembrandt, or Frank Lloyd Wright.

Each creative genius opens up new dimensions of intellectual endeavor and widens the realm of meaning through which man discovers what he is and what he can do. Artists and intellectuals do not reveal ultimate truths, nor do they ever complete the form-fulfilling task of their disciplines. Their virtue lies in the impetus with which knowledge is propelled forward. Cassirer did not imply that there is any particular goal for the creative act. Neither knowledge nor art has any criterion extrinsic to man by which it can be measured and evaluated. One might just as well ask if there exists a method of evaluating the superiority or inferiority of any coexisting living creatures, whether mosquito, bat, or man. The act of symbolizing is an organic function of man. The products of the symbolic act constitute the external residue of this function. Thus art and knowledge are what they are because of what man is. The only measure of value, of purpose, of direction is what man himself endows them with. Our appreciation, dedication, and commitment to these values of civilization, then, constitute the sole confirmation of the primacy maintained by the creative product in the spectrum of human life.

Thus it is difficult to understand the attempts of the Michigan school of anthropology to measure in a quasi-physicalist and evolutionary manner the level of advance in terms of the quantity of energy available for use by that culture. A practical example might be ancient Rome. Because of its great military, economic, and political power, it would be ranked far ahead of the disorganized and fragmented city-states of Greece and presumably even ahead of the subsequent eras of the Renaissance when a similar state of political and social chaos swept over the Italic peninsula. To Leslie White, the Greek intellectual, political, and esthetic achievements would be subjective and irrelevant measures of the evolutionary progress of mankind.[14] In his use of energy as a quantitative criterion, White feels the need to subject culture to a more basic and substantialistic stratum of objective reality which lies outside of the cultural domain; he does this in spite of his purported claims for a strictly culturological set of principles.

Further, his use of physical or material standards to evaluate evolutionary progress denotes his acceptance of traditional adaptive criteria by which we measure the relative success or failure of a culture to establish itself. What the goal of cultural evolution would be if predicated on the basis of an ever-increasing control and utilization of the energy resources of the universe is hard to conceive. Likewise it is difficult to know whether, at the acme of energic control, man's life would be qualitatively any different from his stone-age ancestors, even though he had greater power at his disposal.

Intellectual innovation and control and a richness and diversity of symbolic envisagement thus become the themes of a critico-idealist philosophy of history. History becomes a study of man's struggle to understand and control his individual energies and his capacity for symbolic expression. If, in addition to riding the currents of evolutionary time, man and culture can be said to be going anywhere, man at each moment in time has the power and authority to decide its direction. Edward Sapir once stated that when the entire spectrum of man's cultural activities is examined there is very little that touches ground. Ground, to Sapir, connoted the *terra firma* of organic adaptation. If this indeed is the case, then an examination of the development of human society and culture must be made through criteria internal to the process in terms of the coherences, tensions, the rate and the quality of innovations and change that might take place in any one era. The philosophy of human nature articulated by Cassirer implies that the more we exploit the inner structural principles of the various modes of symbolic thought with which we are endowed, the more we integrate these meanings into an organic fabric of cultural values, and the more we regularize the process of cultural innovation, the closer we come to participating consciously in this perilous journey.

3

DISCURSIVE KNOWLEDGE

For a final examination of the significance of Cassirer's views on the nature of knowledge and thought let us turn once more to the historical perspective. We have attempted to show that Cassirer's philosophy is a link with the past in epistemology—a past that has been gradually freed from its traditional metaphysical encumbrances. It has been suggested that Cassirer's epistemology is in the main line of contemporary thought, that indeed it acts as a synthesizing agent, bringing together in a potentially significant philosophy of culture a unified position through which we may appraise and understand the various strands of human thought. On the other hand, it is true that Cassirer was less influential as a philosopher on other philosophers, for he certainly deviated from the general analytical concerns of Western philosophy. In terms of the propagation of affiliated schools of philosophy or disciples, Cassirer's philosophy brought forth sparse fruit within the profession.[15]

A number of reasons could account for this lack of influence. 1) Cassirer was at the beginning a member of one of the lesser-known philosophical schools in Europe. The Marburg Kantians, in upholding their special historical insight in the theory of knowledge, were in a nebulous position, to the left of the more ambitious *geisteswissenschaftlichen* idealist schools and to the right of the more radical positivist groups that were springing up at the end of the nineteenth century. 2) In attempting to use history as a vehicle of understanding the present and in interpreting then current dogmas in the theory of knowledge in terms of past achievements in philosophy, Cassirer broke sharply with the current antihistorical tendencies in philosophy. 3) His detailed contextual analysis of the various empirical domains, in culture as well as in physical science, has diverted his influence from the dominant methodological, linguistic, and logical concerns of modern philosophy. As a consequence, Cassirer has been influential in areas as diverse as comparative mythology, history, and physics. However the sum total of these influences has not been enough to attract a significant group of followers to his overall position.[16] 4) At a time when the only counterpoint to the dominant analytical trends has been philosophers in search of a way of life, political behavior, and religious salvation—resulting in existentialism, Marxism, neo-Thomism, and Protestant idealism—Cassirer's classical concerns with epistemology, combined with his rather austere continental style, have possibly lessened interest in his writings.

This latter point can best be exemplified in Philipp Frank's analysis of the neo-Kantian movement, especially Cassirer's epistemology. In spite of Frank's somewhat jaundiced view of what neo-Kantianism stood for at the turn of the twentieth century, his analysis of Cassirer's *Determinism and Indeterminism in Modern Physics* was both detailed and forthright. Apparently Frank recognized that nothing existed in this book to challenge the empirical and presumed antimetaphysical nature of his logical positivism. Yet something in Cassirer's language led Frank to suspect that there was a covert metaphysics lurking behind Cassirer's exemplary analysis of quantum theory. It was so strong a suspicion that Frank rejected Cassirer and *Determinism* as in effect representing a philosophical atavism that had best be relegated to the earlier century and forgotten.[17]

4

It is not surprising that Cassirer appeared to be a holdover from an older and reactionary tradition. Frank was a representative of a more contemporary philosophical movement. Nevertheless, it is an ironic evaluation, for it stems from Frank's inability to appreciate Cassirer's broader philosophical position. No veiled metaphysic lurks behind Cassirer's theory of scientific knowledge, as had been the case with Pierre Duhem's covert Aristotelianism.

As a matter of fact, Frank's accusation is doubly misplaced. For Frank himself represents the overly confident new generation that, as Cassirer pointed out, is itself enmeshed in the historical dialectic that regularly overtakes the theory of knowledge. With every great scientific revolution has come inevitably a loss of an instrumentalist and functional understanding of the structure of knowledge. And it is the younger generation which, in its enthusiasm to embrace the new knowledge, to understand its meaning and discourse on its principles, begins to enunciate what are subsequently discovered to be metaphysical statements.

This ought not lead to handwringing. The predictable debates within the new tradition will ultimately bring forward a more viably critical analysis of the status of knowledge. Logical positivism, Frank's own movement, fell into the same philosophical trap in its attempt to extirpate the metaphysical thinking of the earlier tradition. Man seems to have a congenital historical weakness for stumbling into such intellectual postures. At any rate, this group too, after much internal debate, recovered the high road of epistemological instrumentalism.

The most important philosophical force in the positivist movement was Ernst Mach. Mach made several significant contributions which helped direct this new group of scientists, mathematicians, and logicians toward an investigation of the philosophical foundations of science, as they had not been studied for over a century. First was Mach's attack on several substantive claims of Newtonian mechanics; one such was absolute space. The arguments he adduced against this concept paved the way for Einstein's reevaluations, later incorporated into the general theory of relativity. In addition, Mach's broadside attacks on the supposedly absolute character of mechanical explanations—at a time when Newtonian science was being undermined by new conceptual structures—were influential in the forming of many of the positivistic views on the nature of scientific theories.

Perhaps most important was Mach's attempt to avoid grounding theory in unobservables and his empiricistic rooting of science in concrete experience, phenomena, and raw sensory data. He felt that if science had this empirical basis, it would provide a firmer foundation on which to graft a more objective and secure structure of knowledge. The fact that Mach avoided attributing higher significance to theories, which were to him merely a descriptive shorthand to correlate the phenomena of experience, seemed to add to the heuristic advance of experimental science.

There were other influences in the positivistic movement as well; the work going on in the foundations of mathematics was of special significance. Whitehead and Russell, starting in 1910, had published the *Principia Mathematica*, a work which explained mathematics in terms of symbolic logic. Russell himself had been deeply influenced by the work of Georg Cantor, Giuseppe Peano, and especially Gottlob Frege (his analysis of the nature of numbers). In addition, David Hilbert's axiomatization of the foundations of geometry offered a logico-deductive model that could be used in the systematization of theoretical physics and ultimately to the philosophical analysis of the logical foundations of science itself. The union of mathematics, physics, and philosophy was an organizing tenet and preoccupation of this school of thought. Its object was to purge philosophy of traditional metaphysics and to bring about a proper methodological approach in the establishment of objective truth, truth that would not be limited to any one particular state in the development of scientific knowledge.

In creating such a philosophy, it was important first to establish this enterprise on a firm foundation. Perhaps the most problematic issue was one which Ernst Mach had left somewhat open. This consisted in his empiricistic and nominalistic views on scientific theory. The various papers by Einstein on relativity had had a great impact on the logical positivists. For if Mach's strictures against mechanics and Newton's absolutes had borne creative scientific fruit, the particular theories which had been enunciated as a consequence bore little resemblance, in their highly abstract character, to Mach's requirements of the close correlation of theory with perceptual experience. Thus a gap had been created between those indubitable requirements for an empirical and concrete source of scientific theory, our sensations, and the logical and highly abstract conceptual structure of physics. A system of knowledge was needed by which one could traverse the various realms of concrete factual experience and the abstract and theoretical extrapolations of physics without falling into the traditional metaphysical error of postulating unverifiable and hypothetical entities and substances (noumena).

The most immediately influential philosophers of the positivistic tradition were Bertrand Russell and his student Ludwig Wittgenstein. Russell was a primary inspiration for the continental interest in the relationship of mathematics and logic to the new philosophy. Russell's solution to the problem of the sources of scientific knowledge and its consequent status was to postulate that all knowledge originated in hard or noninferred data—essentially sensation, whether perceived as inner or outer—as distinguished from common-sense objects and scientific knowledge. This data, which he broadened to include logical truths as well, had to "resist the solvent influence of critical reflection."[18]

Science became a logical construct of events which we ourselves have inferentially ordered after receiving the ultimate and irreducible perceptions of experience. On a level of reality, objectivity, and facticity, then, there is nothing prior to sensation, either experientially or logically. The world of knowledge is parallel to the world of perception. Knowledge is immediately constructed from these sense data through rules of inference supplied by the basic rules of logic. There is thus an infallibility in the use of logic to translate perception into objects and objects into constructs, just as our language translates bits and pieces of scribbly blue lines into a structure of ideas and thought. To account for this translation, therefore, we must admit that there is a correspondence of thought and things, of percept and concept, that the translation is not a mere convention but partakes of something essential in the external state of things. In short, we must have a correspondence theory of knowledge, which would therefore imply the existence of a real context within which this correspondence takes place.

Wittgenstein, whose *Tractatus Logico-Philosophicus* was published in 1922—and studied devotedly by the members of the youthful *Wiener Kreis*[19]—adopted the general stance of Russell. Not only did he accept Russell's universe of ultimately real entities and perceptions and his copy view of language and reality. He also accepted a view of the derivative character of propositional and thus scientific knowledge.

In neither philosopher do we find the scientific realism that bemused nineteenth-century scientific philosophers. But we do find correspondence theories of knowledge (demolished almost a century earlier by, among others, Helmholtz and Hertz), the necessity to look for the origins of knowledge in an ultimate category of the real, and perceptions or sensations of the kind which haunted Mach's metaphysics. Also, perhaps as a saving grace,

Wittgenstein and Russell represented scientific knowledge as a construct of logic, reflection, and thought. This was an element of Duhem and seemed to transcend the limitations of their commitment to Mach. This idealist element in the status of scientific theory may have been due to Russell's logical and mathematical background as well as to his early interest in the philosophy of Leibniz.

Wittgenstein differs from Russell in one other important intellectual respect. While he did not have the latter's almost infinite flexibility in adjusting his views to criticism, he was able to repudiate the epistemological position developed in the *Tractatus* and adopt a completely new position. In the *Tractatus*, Wittgenstein postulates that knowledge originates in the realm of propositional statements. These mediated propositions are, however, pictures of reals, of which there are underlying atomic facts—*Sachverhalt*. The atomic facts constitute a realm of infinite possibilities which become realized possibilities when the propositions that mirror them are true. If untrue, the proposition is unable to realize the atomic facts. But this does not render their reality less significant.

Russell views the character of the atomic facts in a similar manner. Here they are perhaps more clearly sensational. The facts become the irreducibles of experience. Knowledge is erected through true propositions based on the union of atomic facts, again themselves situated in a realm of possibility. The structure of knowledge seems to be of a different logical character than these atomic facts. Yet the atomic facts merge to form an empirically united and predictable fabric of experience. This existential compatibility, Wittgenstein surmises, is based on a correspondence, a mirroring of each other so that diversity somehow becomes unity. But this must occur in the context of another realm of experience, in the world beyond appearance. Thus Wittgenstein's even more intense realism comes through. The ultimate ontological reality is located in fact or sensation, yet in the marvelous correspondence of particular and general, of percept and concept or proposition, an inner likeness is in some way predicated. We can only conclude that this likeness must occur in a reality outside of experience.[20]

While the logical positivists as a group were in a sense coalesced by Wittgenstein's *Tractatus*, Moritz Schlick, an original member of the group, had already published a work on the foundations of epistemology which in a number of ways anticipated Wittgenstein's ideas.[21] From the time that Schlick arrived in Vienna to assume his chair in philosophy in 1922 until the end of that decade, it was his and Wittgenstein's influence that directed the project of guiding philosophy away from the excrescences of metaphysics onto the safe road of scientific and logical pursuits.

To attain scientific indubitability, however, according to Schlick and his philosophical ally, Friedrich Waismann, we must look to the experiential, thus the real, not to the hypothetical and unverifiable entities. This nonhypothetical and irreducible level of experience is formed from perceptual propositions, *Konstatierungen*, whether they occur at the onset of experience in the act of building a structure of knowledge or whether they constitute "ultimate elements in terms of which verification is made."[22] Insofar as scientific truth is constructed, the conceptual and symbolic nature of theory is here again appreciated. Still these theories must be anchored to sensational reality, the so-called atomic facts. "Every theory is composed of a network of conceptions and judgments, and is *correct* or *true* if the system of judgments indicates the world of facts *uniquely*. . . . It is, however, possible to indicate identically the *same* set of facts by means of *various* systems of judgments; and consequently there can be various theories in which the criterion of truth [unique correspondence] is equally well satisfied. . . . They are merely different systems of symbols, which are allocated to the same objective reality."[23] Knowledge is essentially created out of a correspondence between this network of theories and concepts, the symbols of knowledge, and the ultimate objective facts of reality.

It was not long before the steady drift of logical positivism into traditional sensationalist and realist forms of metaphysics was recognized by those both within and without this intellectual community. It was Otto Neurath, who had consistently opposed the metaphysical views of Schlick and Wittgenstein, who helped shift the movement away from these dominant motifs.[24] Neurath was later joined in this effort by Rudolf Carnap. Rather than in perceptions or sensations, the protocols of knowledge were to be located in the propositions with which we describe an essentially incommunicable and private sensory experience. Neurath argued that because of its alleged solipsistic quality, sensory experience had no relevance as an epistemological base for science. Philipp Frank and Karl Popper generally accepted this critique and Neurath's advocated shift toward viewing language or discourse as the foundation upon which knowledge could be built.

Eventually this shift in outlook was accepted as official by the movement. Again it was Neurath, this most consistent maverick among the logical positivists, who led the way with his adoption of physicalism. Philipp Frank explains the shift as follows:

This transition from a quasi-idealistic to a quasi-materialistic language which took place in our group about 1930, has been misunderstood by great many authors. They interpret it as a sudden jump into an opposite type of philosophy. As a matter of fact, the "jump" was an expression of our firm belief that the difference between an idealistic and a materialistic system is logically and scientifically of little importance and that there is actually only a difference of emphasis. The choice is determined largely by the emotional connotations or, in other words, by the language in which the patterns of our general culture is usually described.[25]

Of course this is a statement from within the movement. And while Frank himself is not among those accused of metaphysical thinking, it is understandably difficult for him not to overlook some of these weaknesses. Essentially Frank is saying that since the intentions of the members of the positivistic movement were empirical, scientific, and antimetaphysical, traditional metaphysical statements should not be attributed to them. Critics, therefore, he implied, ought not to examine their philosophical statements per se, but ought to consider their intellectual propensities in the context of the ideal toward which they were striving.

At about the time this shift was occurring, Wittgenstein returned to Cambridge (1929) and published the last work that represented his allegiance to the Frege-Russell and *Tractatus* tradition. In this paper, entitled "Logical Form," he noted the ideal of philosophy as being the construction of a language that would reveal and describe the atomic propositions of reality.[26] The year after his appointment to Cambridge (1930), his philosophical orientation began a turn which in the end culminated in his renunciation of Russellian logical atomism and an absorption with the ordinary language analysis reflected in his *Philosophical Investigations.*[27]

Abandoning his attempt to construct objective language systems based in an absolute and atomic realm of sense data, Wittgenstein transferred his attention to the subject matter of knowledge—language. Rather than view language as mirroring the real, or interpret it as a universal basis for the construction of all scientific knowledge, Wittgenstein adopted a more pluralistic and relativistic view of its function. In thus surrendering both metaphysical as well as scientific and objectivistic pretensions, he was now able to describe language as it is actually and ordinarily used. Instead of prescription there is now to be description.

Philosophical puzzlements of the kind which precipitated the positivistic movement itself occur, he claims, because of the bewitchment of our intelligence by faulty use of language. But there is no one remedy. There are as many languages as there are functions in which man engages it. Wittgenstein likens words to tools—there are as many tools as there are uses for them. The different kinds of knowledge men produce are the result of the varying language games that men play. The job of the philosopher, Wittgenstein proposes, is to explicate the rules as they actually function in use. In addition, by keeping constantly alert for possible intellectual abuses of the ground rules of language use, he hopes we can lead the "fly out of the fly bottle."

Even though Wittgenstein's rejection of the scientific concerns of the *Wiener Kreis* was an important loss, his association with its membership had always been essentially indirect. It is necessary to turn back to the Continent to note the general direction of this movement. Symbolic of the logical positivist movement is Rudolf Carnap, partly because of his consistent dedication to its original ideals and partly because of his wide-ranging intellectual interests. He alone pursued these ideals through their various philosophical permutations and most clearly represented the dialectical evolution of this movement. Carnap early supported Schlick in his empiricist view that metaphysics be excluded from science and philosophy, and that as a result we must base knowledge carefully and precisely in sensory experience. The primary therapeutic concern of this movement necessitated the exclusion of any unobservable philosophical constructs.

Carnap's first important work, *Der logische Aufbau der Welt* (1928), though still showing an allegiance to Schlick's views on knowledge, already manifests certain subtle differences which perhaps reflect the results of the dialogue between Neurath's physicalist and somewhat conventionalist views on knowledge and Schlick's more traditional, if informal, empiricism. In this work Carnap displays a more rigorous logical approach than Schlick, one which was to extend itself more systematically in his later writings. This slight modification of his generally phenomenalistic orientation lay in his rooting of the primitives of knowledge in cross sections of experience—which he called "quality classes"—from which more discrete sensory classes (color, heat, tone) were to be extrapolated. In this analysis he gives tacit recognition to the fact that these so-called sensory classes are not in themselves the primary elements of experience but are logically extracted through the recognition of similarity and dissimilarity from the "quality classes."[28]

In two subsequent articles, written in 1932 and published in *Erkenntnis*, one on the "Unity of Science" and the second on "Protocol Statements," we see Carnap shifting away from his earlier phenomenalism, for the most part under the steady prodding of Neurath. Neurath's structural views on scientific knowledge forced Carnap to concede that truth and falsity in science are a product of linguistic or logical analysis. Before matters of verification can be adduced it is necessary to translate the protocol class of observation sentences into a logical form that might be communicable in the building up of further scientific propositions.[29] The philosopher's task in the adjudication of knowledge consists of a logical process of comparison of ideas subsequent to observation. Carnap still avers, however, that the statements of which science consists must be translatable both into their original observational and their primitive protocols. Thus not only is there a constant comparison of ideas on a level of logical discourse, but there is a commutability for each statement from observation to language and back again to observation. This one-to-one correlation is retained as a criterion of what can be admitted into science so that there is continuous contact with the ultimate protocols of experience.[30]

The *Logical Syntax of Language*, published in 1934, was Carnap's most influential book. All his commitments to Mach's sensationalism and nominalism in scientific verification of theory were now withdrawn. Carnap's understanding of the inner structural character of theories of high generality and the manner in which scientists have developed and utilized theories without themselves necessarily being drawn into metaphysical entrapments had apparently convinced him that the stipulation of a one-to-one correspondence between the symbolism of science and an observational language was an artificial and superfluous methodological constraint.[31] Indeed, he came to recognize, as did Duhem before him, that scientific principles themselves were never adopted or abandoned on the basis of any single disconfirming experiment or observation. In agreement with most scientists, Carnap now felt that the only real necessity in the structure of physical theory was that at some point in the system of laws and principles, reference to observation and experiment would become necessary. But this requirement could never be a fixed or rigid absolute. It had to depend upon the specific context of the theory.

The dissolution of a succession of foundation points upon which one could establish a science of philosophy and thence an objective structure of knowledge thus led Carnap to a new approach, which was elucidated in this volume. Carnap's study with Frege and his subsequent intense interest in logic and mathematics, especially Hilbert's axiomatic method, now led him to look for a firmer anchoring of philosophy and science in logic. If one could not find a universally valid cognitive criterion in experience and sensation, perhaps one could establish it in the form of science rather than in its content.

The axiomatic method was developed through Hilbert's metamathematical researches. It was developed with the hope of banishing from the logic of mathematics all references to an external world of things. Mathematics referred to no external reality; it was a discipline whose subject matter consisted in drawing certain necessary conclusions on the basis of a given set of axioms or postulates. The inferences drawn in the creation of mathematical truths (theorems) were based on the transformation rules established within the system of implicitly defined logical symbols. These truths therefore did not depend on the exigencies occurring in the domain of experience or to any content to which they might be assigned as a means of clarification and explication. It was hoped that the entire system of mathematics might be subsumed under this method to lay bare the inner logical connections of the elements, the nature of its vocabulary, the combination rules by which the elements were to be joined into larger groups of ideas and sentences, and in turn how these sentences were to be translated into and substituted for each other.

This was a completely logical ideal, encompassing an arbitrary set of symbols having no meaning outside this particular domain of ideas. Hilbert and other logicians and mathematicians hoped that each branch of mathematics might someday be developed as consistent and completely axiomatic systems. Eventually, the entire domain of mathematics might be unified and subsumed under these canons.[32] Analogically applied from mathematics to the world of things and people—an endeavor which essentially lay outside the concerns of these mathematicians—this model might prove to be a standard of thought by which we could evaluate the achievements of every science. Thus the more completely the content of science could be subsumed to the logical structure of axiomatization, the closer it would come to the ideals of internal rigor and rational control. Not concerned with any particular domain or subject matter of science, this method would then constitute a universally applicable formal yardstick by which to measure the progress of knowledge and truth.

This in sum was what Carnap began to strive for.[33] He was no longer interested in the language or content of science. He was interested rather in the logical syntax of science, its operative rules or grammar. By centering his attention on the "logic of science," he would shunt the issue of establishing the truth or falsity of propositions to scientists within the disciplines. The work of the philosopher would henceforth be to establish for science the formal or general rules that would facilitate the forming of the proper conclusions, on the basis of any given set of axioms. Wherever problems of theory construction might henceforth intrude, and where within any given discipline of science it might become necessary to translate these formal rules into the particular language of that science, philosophy would play its role.[34]

It was Carnap's view that given a rich enough language it was possible to express the syntax or logical structure from a standpoint within the language itself, so that it might, so to speak, judge itself. In this he hoped to avoid a correspondence view of knowledge or the postulation of a hierarchical set of languages, each explaining the other *ad infinitum*. It is not, then, the primary concern of philosophy to establish either the direction of scientific investigation nor the truth of scientific statements. Philosophy's task is to decide the form or logical structure of what shall become part of the corpus of knowledge. Philosophy is thus to be the logic of science.[35] We will find objectivity in the paradigm of analytic truth, truth that is abstract, general, and all-encompassing, which will stand as a goal for every empirical intellectual endeavor.

This leads us to a final dimension of Carnap's program. According to Carnap, it is the physical sciences that come closest to the deductive and axiomatic canons that he established as the paradigm of knowledge. The world of material objects moving in a spatio-temporal continuum constitutes that realm which most nearly approaches in its logical rigor (its use of mathematics) our ideal. Thus the philosophy of physicalism, originally urged upon Carnap by Neurath, emerged once more, this time as a general example for the future development of all sciences.[36] Carnap at that time was the most eminent enunciator of physicalism.[37]

Carnap was candid about his own change in views. "In former explanations of physicalism we used to refer to the physical language as a basis of the whole language of science. It now seems to me that what we really had in mind as such a basis was rather the thing language or, even more narrowly, the observable predicates of the thing language."[38] His present position can be stated thus: "The so-called thesis of *physicalism* asserts that every term of the language of science—including besides the physical language those sub-languages which are used in biology, in psychology, and in social science—is reducible to terms of the physical language. . . . We may assert reducibility of the terms, but not—as was done in our former publications—definability of the terms and hence translatability of the sentences."[39]

Physicalism was intertwined in a still more general ideal, also in tune with the aspirations expressed through the axiomatic method—the unity of science. Those who advocated the physicalist and "unity of science" theses did not imply that all science would one day have to adopt the vocabulary of the physical sciences or define their own particular content and laws in terms of those traditionally expressible in physics, e.g., force, energy, atoms, relativistic mass, quantum of energy, etc. Rather it was the hope and expectation that all scientific languages would one day have the rigor and deducibility which are now present in physics. Whatever the future evolution of the particular "thing regularities" and "experimental invariances" might be in any discipline, it was hoped that they would present a formal contour, a syntax having the necessary transformational rules that could bridge the chasm that seems at present to separate so many intellectual domains. Thus, here too the concern became methodological. It was hoped that the stipulation of logical canons of truth might establish a criterion by which we could admit any subject matter area into the halls of science.[40]

The difficulties in the "unity of science" program were also many. Even Herbert Feigl, one of the founders of the movement, raised questions as to the hope for such a "radical reduction":

> The thesis which I wish to clarify and whose present value I wish to appraise is . . . *physicalism in the strict sense*, postulating the potential derivability of *all* scientific laws from the laws of physics. . . . As it is an open question as to whether biology, psychology, and the social sciences are ultimately reducible to physical theory we cannot afford to be dogmatic—one way or the other. It is also expressly admitted that speculation in the direction of this sort of unification may in many fields be premature and therefore possibly harmful to scientific progress.[41]

5

But even this considerable modification of the positivist's position—what could be considered a hesitant retreat into a methodological and formalistic objectivism and realism—was subject to the vicissitudes of critical analysis. The basis for questioning Carnap's later program came from a metamathematical study undertaken within the ranks of the inner group itself. In 1931, Kurt Gödel, then a member of the Vienna Circle, published a paper entitled "On Formally Undecidable Propositions of *Principia Mathematica* and Related Systems."[42] Because of its great abstractness and complexity, this important paper did not receive widespread notice for a number of years.[43] It was addressed to the program expressed in Whitehead and Russell's *Principia*, but also to the attempts by Hilbert and others to lay down a program for the complete axiomatization of the various domains of mathematics.

Gödel pointed out that these programs themselves had developed certain internal contradictions which made their achievement doubtful. First, he argued that no metamathematical proof was possible that could prove the formal consistency of a system complete enough to contain the whole of arithmetic, or even systems as simple as the arithmetic of cardinal numbers, unless the metamathematical proof employed rules of inference that were much more powerful than are the transformation rules used in deriving theorems within the system.[44] Thus the consistency of a system such as *Principia Mathematica* could not be demonstrated through the system's own internal resources. It must develop a metamathematical language richer and more complex than that of the *Principia* itself.[45] As Nagel and Newman point out, one dragon is slain only to create another.[46]

Secondly, Gödel concluded that all formal logical systems which might expect to develop and exhaust the possibilities contained in a mathematical domain such as arithmetic were themselves essentially incomplete. Thus given any consistent set of arithmetical axioms there are true arithmetical statements not derivable from the original axioms.[47]

The implications of Gödel's conclusions were far-reaching, but as Nagel and Newman point out, they are not as yet fully understood.

> They seem to show that the hope of finding an absolute proof of consistency for any deductive system in which the whole of arithmetic is expressible cannot be realized, if such a proof must satisfy the finitistic requirements of Hilbert's original program. They also show that there is an endless number of true arithmetical statements which cannot be formally deduced from any specified set of axioms in accordance with a closed set of rules of inference. It follows, therefore, that an axiomatic approach to number theory, for example, cannot exhaust the domain of arithmetic truth and that mathematical proof does not coincide with the exploitation of a formalized axiomatic method.[48]

They seemed to transform all supposedly and hopefully closed systems of proof into open-ended structures. They also showed that the human mind as a creative instrument of thought could not be limited by any fixed system of axioms, themselves produced by thought. The brain, if conceived as a calculating machine, turns out to be a machine far more powerful, complex, and subtle than any nonliving computer or robot yet envisaged.[49] Thus one might conclude that the human brain cannot restrict thought against its own possibilities.

It would seem that Carnap's logical efforts were undermined on the evidence of a member of his own school of philosophy. For, if these metamathematical aspirations are ruled out for a discipline as delimited in meaning as mathematics, then it would hardly be possible to argue that any empirical science—physics or other—could have all its theorems formally deducible from its axioms or, on the other hand, be able to show the consistency of the logical system internal to itself. No science henceforth can be measured against an anterior criterion of consistency or completeness. Every science must now be endowed with a measure of logical indeterminacy and openness that forestalls future attempts to state unequivocally that one principle is meaningful and another meaningless on the basis of a comparison with the given logical syntax of the theoretical system.[50]

The situation with regard to the physicalist hypothesis is less clear-cut. But here too there are important theoretical grounds for questioning the tenability of these methodological aspirations both with regard to the form of all scientific theories and to the expectations that are encompassed in "unity of science" movements. This of course is not to deny that the world, inorganic and organic, cultural and physical, is constituted of matter and conforms to the laws of physics. The question remains however as to whether both organic and cultural phenomena, when analyzed in functional terms, can elicit principles which will be referable in any meaningful way to the form or content of physical principles. Surely the more we learn about molecular interactions the more we will learn about biological processes. And certainly the more we know about biological processes, the greater will be our awareness of the underlying dimensions of culture.

Erwin Schrödinger, for example, in his justly famous *What Is Life?* has made us very aware of the tenuous line which separates even such disparate realms as quantum theory and genetics.[51] Thus what we usually think of in terms of purely biological actions—for instance, a gene mutation—he has shown to be conceptually translatable as a product of a quantum jump within the atom itself. While we are aware that physico-chemical activities are structurally at the root of biological events, it is still unclear whether it would be either possible to view the evolutionary process as a product of mutational change in terms of quantum actions in the genoplasm or fruitful to understand the dynamics of biological interactions from this standpoint, since these processes are so difficult to isolate even in the experimental and laboratory sense and thus would not be rendered meaningful outside the natural context.[52]

The view of physics as the central discipline with which all other disciplines should be correlated has been seriously questioned as a theoretical possibility in a recent book by Walter Elsasser.[53] Elsasser contends that recent advances in knowledge do not further the methodological program of the unity of science; on the contrary, they inhibit any further commitment to its aspirations. He states that the realm of biological processes is inherently impermeable to the reductive or unifying embrace of physics. This impermeability is more than a matter of complexity, large numbers, and statistical difficulties.

If we were considering the program of physicalism in traditional Newtonian terms, there might have been "theory of error" questions raised with regard to the behavior of an entity such as a molecule because of smallness of size, numbers involved, and the consequent particular statistical and computational problems generated. At this time, however, our conceptual difficulties are a consequence of what we know about the behavior of molecules, atoms, and electrons. Also with the advent of quantum theory it is in principle impossible to determine simultaneously those two fundamental aspects of Newtonian determinism—the momentum and position of a particle.

So too, Elsasser argues, our understanding of the very nature of the physico-chemical structure of life prevents us from predicting what a set of atoms will do within, let us say, a gene. The problems of investigating with any causal specificity the "state" conditions within any one particular living structure, with its millions of cells and almost infinite number of atoms and molecules, are increased to a geometrical component far beyond the most complex congeries of particles in contemporary quantum physics. Therefore, Elsasser believes that, to all intents and purposes, living systems must be conceived as open systems, as self- and inner-directing and not strictly reducible in a deterministic manner to the traditional predictive canons associated with physico-chemical theories.

Niels Bohr, also reflecting on the incommensurability of the two systems, speaks of the cognitive complementarity of biology and physics as epitomizing on a different level the complementarity to be found within the world of matter, energy, and light.[54] In this case it is not that experimental conditions preclude our determining all the necessary information about both areas of thought that would lead to what might be called a complete description within these cognitive domains. The complementarity occurs because even if there were no theoretical barriers to the interpenetration of the disciplines by each other, or at the most it was possible to explain biological processes in terms of physico-chemical laws, these explanations would only partially satisfy us.

Taken by themselves, these physico-chemical laws would have to refer to biological functions in terms of particles in motion, chemical exchanges, etc., whereas biological processes, reproduction, alteration and change, adaptation and natural selection, behavior and personality are intellectual constructs having an intrinsic interest to ourselves. They give us a functional kind of knowledge in this domain which has its own particular texture of laws. The character of biological processes would be irrevocably altered if considered solely from the standpoint of the laws of physics. Therefore it is our attitude toward what we need to know about events and processes that occur in the biological realm that serves to define the kinds of laws that we accept as biological knowledge. It is this implicit psycho-logical requirement of thought that suggests the essentially complementary organization of knowledge into the great symbolic divisions of experience.

Thus, practically every cornerstone that logical positivism sought to establish for its scientific philosophy was severely questioned and in effect sheared from its intellectual structure. First sensationalism and phenomenalism, then the plans for an atomic language of protocol sentences based on perception, were abandoned. In the final stages the logistic foundations for a syntax of scientific theory were undermined along with the expectation that physicalism and the "unity of science" might bring all knowledge into one intellectual structure if not—in their highest expectations—reduce it to one content (physics).

In the end, the intellectual dialectic within positivism concluded with the following results: Knowledge is not reducible to any materialistic or sensationalistic substratum. It is rooted in discourse, in effect, in symbolic communication. And while knowledge cannot be anchored in matter or perception, it also cannot be measured or evaluated by a fixed objective and absolute logical grammar that it must universally mirror. In addition, science has various theoretical structures, each with characteristics that are incommensurate with the logical form and content of other disciplines.

What we have returned to here in the theory of knowledge is essentially the neo-Kantian position of Ernst Cassirer, a position which was first enunciated in *Substance and Function* (1910), outlined as a program for all the fields of knowledge in *Einstein's Theory of Relativity* (1921), several years before the members of the Vienna Circle began their meetings, and carried forward consistently in each of Cassirer's works until his death in 1945.

6

Assuming the above arguments concerning the character of discursive knowledge and the intellectual and historical context in which we have placed the spectrum of human behavior, we can ask where thought is leading knowledge and what its future direction and form will be. In recent years, as objectivistic and reductionist views of scientific knowledge have waned, an appreciation has developed for the history of science—partly as a consequence of the ahistorical concerns of the prior positivistic tradition. It has resulted in a number of interesting books concerned with these questions.

Those having special philosophical importance are the works of Stephen Toulmin and Thomas Kuhn.[55] Kuhn's book, because of its radical thesis and significance for the critico-idealistic position that Cassirer espoused, is especially important. In his *The Structure of Scientific Revolutions*, Kuhn attempts to record and analyze the actual process through which large-scale scientific movements are created, why they flourish while others dissolve. He is not concerned with logical or structural analysis. Rather his book reflects an internal analysis of the interaction of predicament and persuasion as the scientists themselves become party to the acceptance and rejection of their intellectual traditions. Kuhn's philosophical orientation is essentially Kantian, as is evidenced by his explicit indebtedness to Emile Meyerson, Emil Boutroux, Jean Piaget, Benjamin Whorf, and Jerome Bruner. In spite of this admitted bias, his essay, buttressed as it is by a wealth of factual materials, is important in that it argues for the synthetic role of thought and action in an empirical area of intellectual endeavor that has in the main been untilled.

Kuhn attempts to show that our normative assumptions as to how theories may be accepted into the corpus of science, i.e., logical syntax, experimental verification, prediction, probabilistic levels of confirmation, are not supported by the historical facts associated with their acceptance. In this aspect he is in agreement with Toulmin. At the most, Kuhn claims, these external and objectivistic criteria assume a peripheral role in the so-called advance of science. Man is not advancing toward truth through science nor does he stand on the ever-expanding threshold of objectivity.[56]

All great theories of science, Kuhn argues, have developed because scientists have first accepted a pattern of thought which has led them to construct a particular vision of reality. These patterns or models of thought he calls paradigms of knowledge. In the acceptance of any particular paradigm, whether it be Copernicus' astronomy, Newton's mechanics, Lavoisier's chemistry, or Einstein's relativity, an entirely new map of experience is drawn which includes directions for all further map making. In accepting the paradigm and learning to think within its boundaries the scientist acquires theory, methods, and standards which usually constitute an entirely new set of criteria for deciding what the real versus the spurious problems of the discipline are. These new criteria also elicit new solutions to old problems within the science.[57]

The shift from one paradigm to another can take place only when a gradual disenchantment occurs with the older general theory's ability to correlate the facts of experience with the existing explanatory structure. The decline of the older theoretical center is marked by the recurrence of differing explanatory anomalies. The old theory tries to account for these anomalies through auxiliary hypotheses and explanations. But the multiplication of these hypotheses, given the natural inclinations of thought to apply "Occam's razor," initiates an intellectual crisis. The inevitably ad hoc nature of these adjustments eventually distorts the paradigm to the point that it begins to take on a monstrous visage.[58] The periods in science immediately prior to the work of Copernicus and Einstein illustrate this crisis stage of scientific history.

The challenge to a crumbling intellectual structure of ideas is soon met by a series of creative responses. New models, perhaps inchoate at first, are haltingly advanced. Then, usually through the work of a great scientific innovator such as Newton, Lavoisier, Einstein, or Heisenberg, a new construction emerges. The ways of looking at the entire range of evidential materials accumulated from tradition changes. The change is of such a radical nature that no principle is beyond reevaluation.[59] The methods, materials, and constructs of scientific thought are transferred to an entirely new plane. This new stage in knowledge does not represent a cumulative acquisition of truths as some theoreticians have assumed—especially with regard to the relationship of relativity theory to Newtonian mechanics.[60] The new model is of such a radical epistemological nature that standards of truth and verification are incommensurable between models, and scientific supporters of each tend to talk through each other.

Only when one pattern of science begins to dominate—and this occurs, Max Planck once hypothesized, because the older generation finally dies out—can solid and public canons be established as to what constitutes acceptable criteria for the verification of the new science. In certain historical cases the shift was so radical and immediate that proponents of earlier models who refused to alter their views in the face of the new structure of ideas —such as Priestly and Agassiz—were no longer considered scientists.[61] In the end the victory of one set of ideas over another does not necessarily constitute a victory for truth over illusion; it is merely a manifestation of the importance of the new and the innovative to the creative tendencies of each new generation of scientists.

It is at this point of victory that normal science, the science of the textbook and the laboratory, begins to dominate the discipline. This period is characterized by great strides in advancing the logical implications of the new paradigm. It is a period that is calm, comfortable, and prosaic in its workaday attitudes, as scientists investigating the interstices of their disciplines advance a well-articulated and comfortable set of scientific truths. Now men begin to speak of the progress of science. Kuhn notes that it is only relative progress, for it is measured within a world of ideas that are themselves conditioned by the choice of one among a number of possible invariant constructs, whose theoretical and empirical implications are in process of being worked out. Progress in scientific history, Kuhn indicates, is never the kind of intellectual advance established by criteria that lie outside the particular theoretical limits of the dominant paradigm, as in logical positivism. But here too, scientific advance must be slow and the cycle must eventually repeat itself as new anomalies occur; the rigor and beauty of the theory begin to be marred by ancillary appendages. Problems and paradoxes will occur and multiply until once more we need intellectual renewal.

Kuhn does not give us specific instructions for interpreting his theory. What does it mean for the direction of science? Clearly he rejects the objectivistic and realist view that science is advancing unhesitatingly toward the truth, if by truth we mean an ultimate and permanent conception of the method or substance of knowledge. In consonance with Cassirer, Kuhn also seems to reject as unlikely Heisenberg's and Margenau's expectations for a deductive system of science based on the discovery of the ultimate invariants of reality.[62] Any historical set of invariants must be interpreted solely as logical *a priori* conceptions from which the paradigm finds its intellectual *raison d'être;* it cannot be attributed to the ontological nature of things.

The implication here is that any fixed conception of the development and nature of scientific truth will founder on the shoals of contingency. Experience will always bring forth its disconfirmations. But it is never mere experience, because thought is an active, searching, and creative factor which shapes experience, gives it form, meaning, and structure so that paradigms can be created through which and by which the world—scientific and otherwise—is accommodated to human needs. And when thought acting in interaction with a dynamic cultural and natural environment has exploited the truths of any particular theoretical model, it will lay it aside and search for a new foundation point from which knowledge may again be reconstructed.[63]

Can we accept Kuhn's extreme relativistic view of the evolution of science? Has there then been no progress in any of its several senses? We should, on the basis of Cassirer's neo-Kantianism, draw a line here. It cannot be that the history of scientific thought in the last few centuries merely records an aimless drifting in a sea of intellectual convenience. There is a rationale for the common-sense assumptions that science has progressed in and of itself without regard to the technological advances of recent times. In asserting the dictum of progress we are not retreating to orthodox absolutistic views of scientific truth and objectivity. We merely reiterate what constitutes the results of Cassirer's phenomenological analysis of the development and evolution of discursive knowledge.

If knowledge and science advance, they do so in several senses. First, science has consistently extended its theoretical suzerainty over increasingly greater areas of experience. The progress of science has been marked by a continuing interaction of the theoretical and common-sense dimensions of experience. Thus with each new theoretical advance a new perception of what the "facts" of experience are becomes possible. In turn the demands of thought for unifying theories result in their anomalous and unassimilated aspects. These in turn precipitate the realignment of the theoretical base of science, setting it once more on its perennial hegira.

Thus as scientific theories increase the scope of their concern, as they subject larger areas of experience to law, the possibility for error is concomitantly increased. The function of science is here realized, for science does not complete itself in the task of fixing knowledge in any one particular state of its development. Science, as Kuhn notes, is characterized, indeed fulfilled in its movement, in vaulting from one set of invariances to another. But each of these jumps is taken as the result of the demands of scientific thought to seek a deeper and more solid foundation.

Second, these patterns of theoretical change are unilinear: they always point in one direction. Scientists will rarely surrender as a heuristic intellectual tool a large and universalizing (even if imperfect) theory until they have another just as general within which they may conceptually operate. No matter how satisfactory their discrete experimental laws, no matter that the laws are not capable of being subsumed to the larger theory, the scientist will allow both to coexist. Indeed he will not falter in his perseverance at uniting the two sets of principles in some kind of logical relationship. The later history of mechanics is a good example of this pertinacity by the majority of theoreticians. Even though large experimental domains, such as electromagnetism and thermodynamics, resisted assimilation to the Newtonian program, many contemporary scientists remained loyal to its theoretical goals until Einstein's theory of relativity, at least as general and encompassing, offered an alternate and empirically satisfying choice.

This advance was achieved by leaving the domain of commonsense experience and pictorial models. It was made through mathematical considerations of the highest level of abstractness, which involved, as do many scientific revolutions, the violation of contemporary canons of common-sense thought. And as science continues to advance, at least in the physico-chemical spheres, the generalizing trends of thought will no doubt be facilitated, accompanied, and achieved by means of the language of quantification given in mathematics.

In the broader historical sense there is progress in knowledge to the extent that man strives to realize the inner logic of thought. And if one can say that discursive thought is dominated by principles of universality and ideality, then we should be able to examine the state of knowledge in the various disciplines to evaluate their varying levels of theoretical encompassment. Yet in what sense could we say that progress had taken place? Even using criteria outside the discipline, let us say for example a cultural yardstick of commitment to the pursuit of knowledge as against political, technological, or esthetic concerns, would yield little, since discursive achievement is not limited to any one society but is truly a transnational enterprise of the species. Thus the state of research in one nation can have no long-term significance on the growth of knowledge in general.

Since each of our paradigms must be seen as leading irrevocably toward its own demise, we can be assured that no given condition of knowledge is ever final. We can anticipate, when experimental and experiential investigations cloud our understanding of basic relationships—as they do at present in particle physics—that the time is better than at other periods in the evolution of the discipline for an integration of low-level empirical correlations through a more encompassing theoretical union. Ultimately, however, it is difficult to conceive that experience will refrain from contributing its indigestible fact or event, forever resisting absorption into any unifying Parmenidean synthesis. The situation could be compared to a Zenoistic paradox, in which experience is the escalator forever producing new factual steps, descending toward particularity with just a bit more speed than we ascend the scale of generality.

We are in the dark as to the meaning of this inner drive of thought. No external measures exist by which we can appraise the state and process of knowledge. These trends of thought have occurred in the past; therefore we should be safe in saying that they will occur in the future. We are barred from expressing dark and mysterious thoughts about man's inner nature; we cannot give these expressions the sanctification of being epistemologically meaningful. The only way we can describe human nature is through the symbolic means available to thought. Thus we are caught within a circle from which we cannot break free.

Cassirer constantly reiterated that the process of describing man must be restricted to a delineation of his works. If we retreat from the upper branches of the tree of knowledge, we find parallel branches of discursive knowledge operating from the same logical principles of abstraction and unification but with disparate experiential materials. These branches of knowledge therefore produce distinctive kinds of scientific laws and theories. If we follow these smaller branches back still further to the wider sources of knowledge, we can begin to perceive the other great symbolic forms. Here different logical and psychological valences of thought pursue different symbolic objectives—as expressed in art, religion, technology, and numerous other ideational structures. Now, gaze up into the far reaches of our "tree." Note that the major limbs branch into smaller entities which are so en-twined with each other that from afar it is difficult to distinguish their origins in each of the great divisions of thought—science, art, religion, etc. In much the same way the various dimensions of symbolic thought which we can distinguish after the fact give rise in life to a seamless texture of experience. Outside this con-textual and analogical examination of man's varying creations of thought there is little that we can say about their ultimate signifi-cance that does not ensnare us in our own words.

As a measure of the extent to which absolutistic and objectivis-tic demands for truth might have to be substituted for by a more modest epistemological goal, let us consider the following dilem-mas. First, Kurt Gödel's investigations into the foundations of deductive mathematical systems revealed to us the impossibility of any intellectual system judging itself. In mathematics it was shown that internal consistency and completeness could not be demonstrated within the whole or any portion of the discipline. If they could not be demonstrated in mathematics, it was even less reasonable to apply these kinds of requirements to an empi-rical science in order to establish a finite set of systematic truths from which all other laws and theories of the science might be formally deducible. The open-ended logical as well as substantive character of the sciences seems to have been given significant confirmation.

Secondly, as Cassirer noted in his *Determinism and Indeter-minism in Modern Physics,* quantum theory has revealed to us a new factor in our researches into nature. As we push our research further and further into the microscopic recesses of the physical world, it becomes increasingly difficult to avoid interfering in or becoming ensnarled with what we are trying to investigate objectively.

Possibly we have reached or are approaching the limits of perceptual objectivity. Our great scientific and technological abilities have brought us full circle, at least in physics, where the need to sense and observe merges with the material we are in-vestigating and it becomes virtually impossible to separate the two. The line between observer and observed has been effaced. This could have been expected on the basis of a critico-idealist view of knowledge. While, as we have noted, the mind orders and shapes experience, it can only do so to the limits of perceptual sensitivity. In building scientific instruments to investigate nature which must intuitively incorporate these perceptual limitations, we could expect that eventually our theoretical demands would outrun our sensory information. Thus objectivity in the Newtonian sense cannot be achieved with quanta of light or electrons. (These canons of nineteenth-century science become regulative or imaginative desires when applied to the content of twentieth-century science.) No matter how deeply our instruments probe into nature, though they far outrun our unaided sensory abilities, they too are inherently limited, for what we can know will ultimately depend on what we can perceive, i.e., what we can transmit from the instruments to our own bodily organs.

As time proceeds, these kinds of epistemological paradoxes will increase and we will become aware, in Paul du Bois-Reymond's words, of the existence of man's "intraphenomenal prison." This will not mean that science will cease. Enclosing the perimeter of man's perceptual capacities through knowledge will only lead knowledge back again upon itself to investigate other logical relationships.

Increasingly our concerns in knowledge will be diverted from physical experience to problems associated with the structural and temporal relationships in the various discursive domains. Possibly there will be greater concern for the methods by which physical processes are transformed into chemical and chemical processes into biological and thence finally into culture. True, with each structural accretion the range of complexity is vastly increased; simple reductive models are inadequate to this problem. Nevertheless, as long as experience demonstrates that culture is a historical product of biology and that biology leads eventually to questions about the basic physical interactions of nature, there will always be a challenge to thought.

The temporal or historical question now becomes even more intriguing. How can we explain the processes by which new layers of complexity have been added to experience? For instance, how did certain complex molecules form themselves into amino acids which later were able to replicate themselves? What were the biological processes by which the brain, through an accretion of structure, was so suddenly enabled to create a set of requirements through which living things might now find fulfillment in something we call culture.

As we witness the inevitable dissolution of the conceptual barriers which divide so many realms of thought, it would be well to remain alert to the easy reductive ensnarements of the past. For example, theoretically we can accept as a valid enterprise the analysis of a violin by the esthetic beauty of tone, quality of wood and varnish, and elegance and form of construction. At the same time the violin can be subjected to acoustical analysis, which would describe it only in terms of its structure of sound. Again, there is no incompatibility in examining a building for its engineering, its architecture, or for its social utility and function. The more varied our symbolic descriptions of any particular experience the richer and more humane our intellectual satisfactions become. Thus physical explanations proposed as the sole paradigm of what can be accepted as knowledge become in fact anti-intellectualistic simplifications and distortions of the real exigencies of our cognitive behavior.

We can look forward in a variety of disciplines to finding this multiple description of experience, what Niels Bohr and J. Robert Oppenheimer have called the principle of complementarity in knowledge. As Oppenheimer has noted, we can expect to find pairs of theories, such as the dynamics of molecular motion and the kinetic theory of gases; the biological description of life alongside a physico-chemical analysis of the same facts; introspective and behavioristic descriptions of thought processes.[64]

There is no basis in fact for defining man in terms which rigidly predetermine our intellectual and social expectations at any future moment in history. The more we know of knowledge the more we will be persuaded as to man's inherently pluralistic potentialities for thought and action. It could be that we have arrived intellectually at the same point that common sense intuits naturally, that man will not suffer any kind of constraint, intellectual or social. The existentialists have attacked philosophy, and rightly, for philosophy has haughtily delimited what man, reality, and thought are and can be. But the anti-intellectualism of the existentialists has less justification. Philosophically we can achieve a commitment to the autonomy of thought and actions without denying, as they do, the role of reason. It is possible for man to function intellectually in a world without having to conform to fixed realities.

It has been hypothesized that the lifetime of a species averages between five and ten million years. The lesser figure probably applies to creatures in the main line of evolutionary advance, such as man. When one considers that it is hardly fifty thousand years since homo sapiens climbed out onto his special evolutionary limb, it is apparent that he is still a youngster among the denizens of this earth. In spite of the fact that he has made his presence known, often with a fatal impact, one dares not imagine the possibilities and direction that his knowledge will take.

Man's knowledge of himself will never be complete or logically sufficient. This we must reluctantly conclude. It is the price we pay for abandoning our traditional and grandiose metaphysical illusions. As Cassirer pointed out, the expectation that we might objectively describe reality is akin to the hope that man could jump over his own shadow. We can only ask that man muster the will to act in the interest of that which his knowledge indicates to be best, for a nature that is ambiguous and ultimately unknown.

Notes

NOTES TO CHAPTER 1

1. E. Cassirer, *The Logic of the Humanities*, trans. C. S. Howe (New Haven: Yale, 1961), p. 48; see also "Some Remarks on the Question of the Originality of the Renaissance," *Journal of the History of Ideas* IV (1943), 51-52.

2. "Some Remarks," p. 52.

3. E. Cassirer, "Galileo: A New Science and a New Spirit," *American Scholar* XII (1942), 7.

4. Cassirer, *The Logic of the Humanities*, p. 49.

5. E. Cassirer, *Substance and Function*, trans. W. C. and M. C. Swabey (New York: Dover, 1953), p. 156.

6. E. Cassirer, *Das Erkenntnisproblem in der Philosophie und Wissenschaft der neueren Zeit*, 2d ed. (Berlin: Bruno Cassirer, 1911), I, 294.

7. On Galileo's redefinition of truth and scientific knowledge, see E. Cassirer, *The Problem of Knowledge*, trans. W. H. Woglem and C. W. Hendel (New Haven: Yale, 1950), IV, 81; also *Das Erkenntnisproblem*, 2d ed., I, 377 ff.

8. E. Cassirer, "Rationalism," *Encyclopaedia Britannica*, 14th ed.

9. E. Cassirer, *The Philosophy of Symbolic Forms*, trans. Ralph Manheim (New Haven: Yale, 1953-57), III, 18-22 (henceforth abbreviated as *P.S.F.*).

10. E. Cassirer, "Galileo's Platonism," in *Studies and Essays in the History of Science and Learning to George Sarton*, ed. M. F. Ashley-Montagu (New York, 1946), p. 280.

11. See Cassirer, *Das Erkenntnisproblem*, 2d ed., I, 380 ff.

12. Cassirer, "Galileo's Platonism," p. 297.

13. See Kant's appreciation of the problem transmitted by Galileo: Immanuel Kant's *Critique of Pure Reason*, trans. Norman Kemp Smith (London: Macmillan, 1958), sec. B xii-B xvii, pp. 19-22.

14. E. Cassirer, *The Philosophy of the Enlightenment*, trans. F. C. A. Koelln and J. P. Pettegrove (Princeton: Princeton Univ., 1951), p. 51; see also *The Problem of Knowledge*, p. 19.

15. E. Cassirer, *An Essay on Man* (New York: Doubleday Anchor, 1953), p. 270; see also *P.S.F.*, III, 454-456.

16. Cassirer, "Rationalism"; see also *P.S.F.*, III, 127.

17. Cassirer, *The Philosophy of the Enlightenment*, p. 13.

18. E. Cassirer, *Determinism and Indeterminism in Modern Physics*, trans. O. T. Benfey (New Haven: Yale, 1956), p. 156 ff.

19. Cassirer, "Galileo: A New Science," pp. 5-6.

20. *P.S.F.*, I, 104.

21. Cassirer, *The Philosophy of the Enlightenment*, p. 51.

22. B. Spinoza, *Tractatus Theologus Politicus*, chap. III, sec. 7, as quoted in ibid., p. 5.

23. *Spinoza's Ethics and "De Intellectus Emendatione,"* trans. A. Boyle (New York: Dutton, 1916), bk. I, proposition XXXIII.

24. E. Cassirer, "Truth," *Encyclopaedia Britannica*, 14th ed.

25. *P.S.F.*, III, 165.

26. Cassirer, *The Philosophy of the Enlightenment*, p. 29. C. Hendel, "Introduction" to *P.S.F.*, I, 23, says that Leibniz brought life back into nature.

27. *P.S.F.*, I, 86.

28. Ibid., p. 112.

29. *P.S.F.*, III, 456-457.

30. Ibid., p. 457.

31. Cassirer, *The Philosophy of the Enlightenment*, p. 114.

32. Cassirer, "Rationalism."

33. Ibid.

34. Cassirer saw this view as compatible with his symbolic view of knowledge, *P.S.F.*, III, 45-46.

35. E. Cassirer, "Newton and Liebniz," *Philosophical Review* XII (1943), 366-391.

36. *P.S.F.*, III, 372.

37. *P.S.F.*, II, 80.

38. Ibid.

39. E. Cassirer, *Einstein's Theory of Relativity*, trans. W. C. and M. C. Swabey (New York: Dover, 1953), pp. 395-398. See also H. Weyl, *Raum, Zeit, Materie; Vorlesungen über allgemeine Relativitätstheorie*, 3d ed. (Berlin, 1920).

40. G. Leibniz, *Mathematische Schriften*, ed. Gerhardt (Berlin-Halle, 1849), IV, 93 ff.

41. *P.S.F.*, III, 205-206.

42. Ibid., pp. 433-434.

43. *P.S.F.*, I, 134.

44. *P.S.F.*, III, 443; see also *Substance and Function*, pp. 330-333.

45. J. Locke, *An Essay Concerning Human Understanding*, bk. IV, chap. 12, sec. 10; see *P.S.F.*, III, 433.

46. *P.S.F.*, I, 136.

47. *P.S.F.*, III, 127.

48. Ibid., p. 290.

49. Ibid., p. 24.

50. Karl Popper, *The Logic of Scientific Discovery* (New York, 1959), p. 19.

51. Cassirer, *An Essay on Man*, p. 65.

52. *P.S.F.*, III, 291. See also *Das Erkenntnisproblem*, 3d ed. (1922), II, 297 ff.

53. *P.S.F.*, I, 139. See also *Das Erkenntnisproblem*, 3d ed., II, 315.

54. Cassirer, *The Philosophy of the Enlightenment*, p. 63.

55. Cassirer, *Substance and Function*, pp. 330-331.

56. *P.S.F.*, I, 101 ff.

57. *P.S.F.*, III, 322.

58. Cassirer paraphrasing D. Hume, *A Treatise of Human Nature*, pt. IV, sec. 2; *P.S.F.*, III, 322.

59. *P.S.F.*, II, 44-46. See also *Determinism and Indeterminism*, p. 17 ff.

60. Cassirer, *Determinism and Indeterminism*, p. 73.

NOTES TO CHAPTER 2

1. For example, the *Neuen Lehrbegriffe der Bewegung und der Ruhe* (1758) and *Versuch, den Begriffe der negativen Grössen in die Weltweisheit einzuführen* (1763).

2. The famous postulation of the nebular hypothesis is given in *Allgemeine Naturgeschichte und Theorie des Himmels* (1755).

3. T. D. Weldon, *Introduction to Kant's Critique of Pure Reason* (Oxford, 1958), p. 9 ff. N. K. Smith, *A Commentary to Kant's "Critique of Pure Reason"* (London, 1930), p. 605.

4. Weldon, pp. 34, 55, 124, speaks of Teton's influence; also see A. Riehl, *Der philosophische Kritizismus* (Leipzig, 1876), p. 209.

5. I. Kant, *Prolegomena*, ed. Lewis W. Beck (New York: Liberal Arts Press, 1951), p. 8.

6. A. C. Ewing, *A Short Commentary on Kant's Critique of Pure Reason* (Chicago, 1938), p. 12.

7. E. Cassirer, *Einstein's Theory of Relativity*, trans. W. C. and M. C. Swabey (New York: Dover, 1953), p. 352. See also on Kant and Euler, E. Cassirer, *Das Erkenntnisproblem in der Philosophie und Wissenschaft der neueren Zeit*, 1st ed. (Berlin: Bruno Cassirer, 1907), II, 698-699.

8. Cassirer, *Einstein's Theory*, p. 411.

9. Gottfried Martin, *Kant's Metaphysic and Theory of Science*, trans. P. G. Lucas (Manchester, 1955). This was discussed in Cassirer, *Das Erkenntnisproblem*, 3d ed. (Berlin: Bruno Cassirer, 1922), II, 586 ff., especially 600-601.

10. N. K. Smith, *A Commentary*, p. 583 ff.

11. *Immanuel Kant's Critique of Pure Reason*, trans. Norman Kemp Smith (London: Macmillan, 1958), B xvi-xvii, pp. 21-22.

12. R. Adamson, "Kant," *Encyclopaedia Britannica*, 11th ed.

13. For a complete discussion, see H. J. Paton, *Kant's Metaphysic of Experience* (London: Allen and Unwin, 1951), I, 73 ff.

14. Ibid., p. 73.

15. See Ewing, *A Short Commentary on Kant's Critique of Pure Reason*.

16. Hans Vaihinger's *Kommenter zu Kritik der reinen Vernunft* (Stuttgart, 1881) has been the source of a great part of this type of criticism. It also seems to have served as the basis for much of Kemp Smith's analysis.

17. C. Hendel, "Introduction" to E. Cassirer, *The Philosophy of Symbolic Forms*, trans. Ralph Manheim (New Haven: Yale, 1953-57) I, 14.

18. See Hendel's discussion of this in Cassirer, *P.S.F.*, I, 9-12. The full treatment by Kant can be found in "The Reduction of the Pure Concepts of Understanding," in *Critique of Pure Reason*, especially A 106-107.

19. *P.S.F.*, III, 193.

20. Adamson, "Kant."

21. Hermann Cohen, "Preface" to 1st ed., *Logik der reinen Erkenntnis*, (1902), xi, xiii, cited by F. Kaufmann, in *The Philosophy of Ernst Cassirer*, ed. P. Schilpp (New York: Tudor, 1958), p. 185.

22. E. Cassirer's *Kants Leben und Lehre* (Berlin: Bruno Cassirer, 1918) delineates Kant's philosophy through his life and intellectual maturation.

23. *P.S.F.*, III, 6-7.

24. Harald Höffding, *A History of Modern Philosophy*, trans. B. E. Meyer (New York: Dover, 1955), II, 60.

25. First edition, A 23-29. See N. K. Smith, *A Commentary*, pp. 625-626.

26. N. K. Smith, *A Commentary*, p. 623, from *Werke*, ed. Hartenstein, I, 457 ff.

27. N. K. Smith, *A Commentary*, p. 617.

28. Ibid.

29. Kaufmann, "Cassirer's Theory of Scientific Knowledge," in *The Philosophy of Ernst Cassirer*, ed. Schilpp.

30. Cassirer, *Einstein's Theory*, esp. pp. 412-413, 449-452; *Determinism and Indeterminism in Modern Physics*, trans. O. T. Benfey (New Haven: Yale, 1956).

31. Cassirer, *Einstein's Theory*, p. 411.

32. Paton, *Kant's Metaphysics*, I, 582, 583-584, n. 1; II, 384.

33. E. Cassirer, "Rationalism"; C. Hendel, "Introduction" to *P.S.F.* I, 12 ff.

34. Ewing, *A Short Commentary on Kant's Critique of Pure Reason*, pp. 145-147.

35. See Paton, *Kant's Metaphysics*, II, 39-41.

36. *I. Kant's Critique*, trans. N. K. Smith, A 138, B 177.

37. Weldon, *Introduction to Kant's Critique of Pure Reason*, p. 164, noted the problem of the subsumption of the particular under the universal to be traceable to the *Parmenides* of Plato.

38. Weldon, *Introduction to Kant's Critique*, p. 162.

39. N. K. Smith, *A Commentary*, p. 335.

40. *P.S.F.*, I, 104.

41. *I. Kant's Critique*, trans. N. K. Smith, A 147, B 186-187.

42. See Weldon, *Introduction to Kant's Critique*, p. 163; Paton, *Kant's Metaphysics*, II, 41.

43. E. Cassirer, *Rousseau, Kant, Goethe*, trans. J. Gutmann, P. Kristeller, and J. Randall, Jr. (New York: Harper, 1963), pp. 88-89.

44. Ibid., pp. 74-75.

45. Cassirer, *Kants Leben und Lehre*, pp. 250-251.

46. Cassirer, *Rousseau, Kant, Goethe*, p. 57.

47. Ibid., p. 23.

48. Ibid., p. 59.

49. *P.S.F.*, I, 79.

NOTES TO CHAPTER 3

1. E. Cassirer, *Rousseau, Kant, Geothe*, trans. J. Gutmann, P. Kristeller, and J. Randall, Jr. (New York: Harper, 1963), pp. 97-98.

2. Überweg, in his famous history of philosophy, compared Thomas's achievement with Kant's. Interestingly, Duns Scotus's role as dogmatist compared with Leibniz's role in the tradition of Kant.

3. See for a full treatment of this: J. T. Merz, *A History of European Thought in the Nineteenth Century* (Edinburgh, 1907), I.

4. A. d'Abro, *The Rise of the New Physics* (New York: Dover, 1951), p. 75; W. P. D. Whiteman, *The Growth of Scientific Ideas* (New Haven, 1953), p. 246.

5. "... given the distribution of the masses and velocities of all the material particles of the universe at any one instant of time, it is theoretically possible to foretell their precise arrangement at any future time." Laplace quoted in Whiteman, *The Growth of Scientific Ideas*, p. 112.

6. Merz, *A History of European Thought*, I, 366; Whiteman, *The Growth of Scientific Ideas*, p. 244.

7. K. Lasswitz, *Geschichte der Atomistik* (Hamburg, 1890), III, 368 ff.; d'Abro, *The Rise of the New Physics*, p. 65; see also Merz's extended treatment in *A History of European Thought*, II.

8. E. Cassirer, *The Problem of Knowledge*, trans. W. H. Woglem and C. W. Hendel (New Haven: Yale, 1950), p. 2 *(Das Erkenntnisproblem*, IV).

9. H. T. Pledge, *Science Since 1500* (New York, 1959), pp. 139-140.

10. H. Hertz, *Untersuchungen über die Ausbreitung der elektrischen Kraft* (Leipzig, 1892), p. 23, in Cassirer, *The Problem of Knowledge*, p. 104.

11. M. Planck, *Das Princip der Erhaltung der Energie* (Leipzig, 1887), p. 136, in *The Philosophy of Symbolic Forms*, trans. Ralph Manheim (New Haven: Yale, 1953-57), III, 462.

12. Ernst Mach, *The Science of Mechanics* (1883).

13. Emile Meyerson's works on the problem of substantiality in physics should not be omitted, e.g., *Identité et Réalité, La Théorie Physique, La Déduction Relativiste*. However, it should be noted that the "cloud chamber" experiments of C. T. R. Wilson and the work with Brownian movement of Perrin in 1910 brought back the "reality" of the atom as an important working scientific construct: Pledge, *Science Since 1500*, pp. 213, 262.

14. E. Cassirer, *Substance and Function*, trans. W. C. and M. C. Swabey (New York: Dover, 1953), p. 119.

15. L. Boltzmann, *Ein Wort der Mathematik an die Energetik* (Leipzig: Populäre Schriften), p. 129 ff., in Cassirer, *Substance and Function*, p. 160.

16. Cf. Pierre Duhem, *The Aim and Structure of Physical Theory*, trans. Philip Weiner (Princeton, 1954).

17. Cassirer, *The Problem of Knowledge*, p. 96.

18. D'Abro, *The Rise of the New Physics*, p. 101.

19. G. Helm, *Die Energetik nach ihrer geschichtlichen Entwicklung* (Leipzig, 1898), quoted in Cassirer, *The Problem of Knowledge*, p. 100.

20. R. Mayer, "Mayer an Griesinger," *Kleinere Schriften und*

Briefe (Stuttgart, 1893), quoted in Cassirer, *The Problem of Knowledge*, p. 99.

21. Cassirer, *The Problem of Knowledge*, pp. 4-5; E. Zeller, as quoted in Cassirer, "Neo-Kantianism," *Encyclopaedia Britannica*, 14th ed.

22. Cassirer, *The Problem of Knowledge*, p. 5.

23. Cassirer, *Determinism and Indeterminism in Modern Physics*, trans. O. T. Benfey (New Haven: Yale, 1956), p. 129.

24. *P.S.F.*, III, 131-134.

25. Cassirer, "Neo-Kantianism."

26. H. L. F. Von Helmholtz, *Handbuch der physiologischen Optik*, quoted in Cassirer, *The Problem of Knowledge*, p. 4.

27. Cassirer, "Neo-Kantianism."

28. E. Zeller, "*Über Bedeutung und Aufgabe der Erkenntnistheorie*," *Vorträge und Abhandlungen* (Leipzig, 1887), quoted in Cassirer, *The Problem of Knowledge*, p. 5.

29. *P.S.F.*, III, 147.

30. Helmholtz, *Optik*, quoted in Cassirer, *Determinism and Indeterminism*, p. 130.

31. *P.S.F.*, III, 147; Helmholtz, *Optik*, as quoted in *P.S.F.*, III, 286.

32. *P.S.F.*, III, 287.

33. Helmholtz, *Optik*, quoted in Cassirer, *P.S.F.*, III, 286-287.

34. *P.S.F.*, III, 305.

35. Ibid., pp. 147-148.

36. Ibid., p. 147.

37. Ibid., p. 146.

38. Cassirer, *Determinism and Indeterminism*, p. 130.

39. *P.S.F.*, I, 168.

40. *P.S.F.*, III, 148.

41. Ibid., p. 147.

42. *Reden* (Ser. 1, Leipzig, 1886), p. 114. See Cassirer, *Determinism and Indeterminism*, p. 7.

43. Quoted in Cassirer, *Determinism and Indeterminism*, p. 6.

44. See ibid., p. 8.

45. *Über die Grundlagen* (Tübingen, 1890), sec. 8, quoted in Cassirer, *Determinism and Indeterminism*, p. 8.

46. A. Rey, *La Théorie de Physique chez les Physiciens Contemporains* (Paris, 1907), p. 16 ff., quoted in Philipp Frank, *Modern Science and Its Philosophy* (New York, 1955), p. 3.

47. Duhem, *The Aim and Structure of Physical Theory*, pp. 333-335. Behind "natural classification" was a modernized Aristotelian metaphysics, through which Duhem would lead science back into the Church.

48. Cassirer, *The Problem of Knowledge*, pp. 94-96.

49. Albert Einstein, *Physikalische Zeitschrift*, XVII (1916), p. 103. Quoted by Philipp Frank, "Einstein, Mach, Logical Positivism," in *Albert Einstein: Philosopher-Scientist*, ed. P. A. Schilpp (New York, 1951), p. 272.

50. *P.S.F.*, III, 410.

51. Ernst Mach, "The Economical Nature of Physics," in *Popular Science Lectures*, trans. Thomas J. McCormack (LaSalle, Ill., 1943), pp. 208-209.

52. Ernst Mach, *Erkenntnis und Irrtum* (Leipzig, 1905), p. 189, quoted in Cassirer, *Determinism and Indeterminism*, p. 83.

53. Ernst Mach, *The Science of Mechanics* (1883), in James Newman, ed., *The World of Mathematics* (New York, 1956), III, 1787-1788.

54. E. Mach, *Die Principien der Wärmelehre* (Leipzig, 1896), p. 422 ff., quoted in Cassirer, *Substance and Function*, pp. 260-261.

55. P. Frank, *Modern Science*, pp. 72-74. Also article "Einstein, Mach, Logical Positivism," in *A. Einstein: Philosopher-Scientist*, ed. Schilpp, p. 271 ff. See also E. Nagel, "Einstein's Philosophy of Science," in *Logic Without Metaphysics* (Glencoe, Ill., 1956), p. 291; for Einstein's reply see *A. Einstein*, ed. Schilpp, pp. 674-680.

56. P. Frank, *Modern Science*, p. 83.

57. E. Nagel, "Einstein's Philosophy of Science," in *Logic Without Metaphysics*, p. 292. See also E. Mach, "The Economical Nature of Physics," in *Popular Science Lectures*, pp. 208-209.

58. *P.S.F.*, III, 30.

59. D'Abro, *The Rise of the New Physics*, p. 95.

60. *P.S.F.*, III, 24.

61. Cassirer, *The Problem of Knowledge*, pp. 94-96; Frank, *Modern Science*, p. 61 ff.; Cassirer, *Determinism and Indeterminism*, p. 140.

62. W. Ostwald, *Grundrisse der allgemeinen Chemie* (1909), quoted in Cassirer, *The Problem of Knowledge*, p. 103.

63. Frank, *Modern Science*, chap. 1.

64. (Berlin, 1928).

65. Frank, *Modern Science*, p. 33.

66. F. S. C. Northrop states that Einstein's conception of science includes the shift from sensationalism to physicalism, in *A. Einstein*, ed. Schilpp, p. 404 ff.

67. A. Einstein in *A. Einstein*, p. 684; also E. Nagel in *Logic Without Metaphysics*, p. 290 ff.

68. *A. Einstein*, pp. 679-680.

69. E. Study, *Die Realistische Weltansicht und die Lehre von Raum* (Brunswick, 1914), cited in Frank, *Modern Science*, p. 65.

70. Study, *Die Realistische Weltansicht*, p. 37, in Frank, *Modern Science*, p. 65.

71. E. Mach, *Die Geschichte und die Wurzel des Satzes von der Erhaltung der Arbeit* (Prague, 1872), p. 31, quoted in Cassirer, *Determinism and Indeterminism*, p. 83, also pp. 68-69.

72. Cassirer, *Determinism and Indeterminism*, p. 83.

73. E. Cassirer, *Einstein's Theory of Relativity*, trans. W. C. and M. C. Swabey (New York: Dover, 1953), pp. 428-429.

74. *P.S.F.*, III, 20. Helmut Kuhn, "Cassirer's Philosophy of Culture," in *The Philosophy of Ernst Cassirer*, ed. P. Schilpp (New York: Tudor, 1958), p. 559, speaks of Cassirer's adaptation of the Hertzian "symbol."

75. Cassirer, *The Problem of Knowledge*, p. 114; *P.S.F.*, I, 76.

76. H. Hertz, *Principles of Mechanics*, cf. *P.S.F.*, III, 20.

77. Cassirer, *The Problem of Knowledge*, p. 104, cites Poincaré's "La Théorie de Maxwell et les Oscillations Hertziennes," *Scientia*, Nov. 1907.

78. Cassirer, *The Problem of Knowledge*, p. 104.

79. Hertz, *Untersuchungen über die Ausbreitung den elektrischen Kraft*, p. 23, in Cassirer, *The Problem of Knowledge*, p. 104.

80. Cassirer, *The Problem of Knowledge*, p. 106.

81. Ibid.

82. Likewise, science is no mere assimilation of facts. Poincaré's famous statement is, "a science is made out of facts, just as a house is made out of stones, but a mere collection of facts is not a science, any more than a pile of stones is a house."

83. H. Poincaré, *The Foundations of Science* (New York, 1913).

84. Poincaré, quoted in Frank, *Modern Science*, pp. 10-11.

85. Cassirer, *The Problem of Knowledge*, pp. 110-111.

86. Frank, *Modern Science*, p. 16; L. DeBroglie, "Introduction" to Duhem's *The Aim and Structure of Physical Theory*, p. ix.

87. Duhem, *The Aim and Structure*, pp. 21-23.

88. *P.S.F.*, III, 21-22, 411.

89. Duhem, *The Aim and Structure*, p. 183 ff.

90. Ibid., pp. xi-xii.

91. Cassirer, *The Problem of Knowledge*, pp. 111-114.

92. Duhem, *The Aim and Structure*, p. 30.

93. The writings of Meyerson figure only modestly in Cassirer's analysis of science, see *Substance and Function*, p. 324. Yet of all the scientific Kantians, Meyerson, in his concern for the phenomenology of thought as expressed in theory, is closest in spirit to Cassirer's epistemological concerns.

94. E. Meyerson, *Identity and Reality*, trans. Kate Lowenberg (London, 1929); original, 1908.

95. H. Poincaré, *Thermodynamique* (Paris, 1909), p. 9, in Meyerson, *Identity and Reality*, p. 209.

96. E. Cassirer, "Neo-Kantianism," *Encyclopaedia Britannica*, 14th ed.

97. E. Cassirer, "Hermann Cohen," *Social Research* X, no. 2 (1943), 217-243; this rejected view is ascribed to Lange and Helmholtz. See Immanuel Kant, *Critique of Pure Reason*, trans. Norman Kemp Smith (London, 1958), B 167-168; B 174-175.

98. Cassirer, "Hermann Cohen," p. 226.

99. Cassirer, "Neo-Kantianism."

100. Ibid.

101. For example, Paul Natorp, "Kant und die Marburger Schule," *Kant Studien* XVIII (1910). Cf. William H. Werkmeister, "Cassirer's Advance Beyond Neo-Kantianism," in *Philosophy of Ernst Cassirer*, ed. Schilpp, pp. 759-798.

102. W. Werkmeister, "Cassirer's Advance."

103. Cassirer, "Neo-Kantianism"; Gottfried Martin, *Kant's Metaphysics and Theory of Science*, trans. P. G. Lucas (Manchester, 1955).

104. I. K. Stephens, "Cassirer's Doctrine of the *A Priori*," in *Philosopy of Ernst Cassirer*, ed. Schilpp, pp. 156-158.

105. Cassirer, *Substance and Function*, p. 269.

106. Felix Kaufmann, "Cassirer's Theory of Scientific Knowledge," in *Philosophy of Ernst Cassirer*, ed. Schilpp, pp. 191-193.

107. E. Nagel, *Sovereign Reason* (Glenco, Ill., 1954), pp. 52-57.

NOTES TO CHAPTER 4

1. E. Cassirer, *Descartes' Kritik der mathematischen und naturwissenschaftlichen Erkenntnis* (1899), used as the introduction to E. Cassirer, *Leibniz' System in seinen wissenschaftlichen Grundlagen* (1902; Hildesheim, Germany, 1962).

2. E. Cassirer, *Das Erkenntnisproblem in der Philosophie und Wissenschaft der neueren Zeit*, 3d ed. (Berlin: Bruno Cassirer, 1922), I, II.

3. E. Cassirer, *Determinism and Indeterminism in Modern Physics*, trans. O. T. Benfey (New Haven: Yale, 1956), p. xxii.

4. See chap. 3, this study.

5. D. Gawronsky, "Ernst Cassirer: His Life and His Work," *The Philosophy of Ernst Cassirer*, ed. P. Schilpp (New York: Tudor, 1958), p. 21.

6. H. Cohen, as quoted in ibid.

7. D. Gawronsky, "Ernst Cassirer," p. 25.

8. H. Reichenbach, "The Philosophical Significance of the Theory of Relativity," in *Albert Einstein, Philosopher-Scientist*, ed. P. A. Schilpp (New York, 1951), p. 299. Also Reichenbach's *Relativitätstheorie und Erkenntnis Apriori* (Berlin, 1920).

9. E. Cassirer, *Einstein's Theory of Relativity*, trans. W. C. and M. C. Swabey (New York: Dover, 1953), p. 400.

10. Ibid., pp. 398, 430 ff.

11. F. Kaufmann, "Cassirer's Theory of Scientific Knowledge," in *The Philosophy of Ernst Cassirer*, ed. P. Schilpp, pp. 185-213, esp. p. 192.

12. Cassirer, *Einstein's Theory*, p. 356. Also A. Einstein, *Geometrie und Erfahrung* (Berlin, 1921).

13. *The Philosophy of Symbolic Forms*, trans. Ralph Manheim (New Haven: Yale, 1953-57), III, 458.

14. Ibid., p. 459.

15. Ibid.

16. A. Einstein, "Reply to Criticisms," in *A. Einstein*, ed. Schilpp, pp. 677-679. Andrew Paul Ushenko in this volume placed Einstein's philosophy somewhere between "Cassirer's Neo-Kantianism and Mach's Positivism," p. 609.

17. Cassirer, *Einstein's Theory*, p. 411.

18. Ibid., p. 412.

19. M. Laue, *Das Relativitätsprincip* (Braunschweig, 1911), as quoted in ibid., pp. 414-415.

20. Cassirer, *Einstein's Theory*, p. 392.

21. Ibid., p. 451.

22. See chap. 3, sec. 4, this study.

23. Cassirer, *Einstein's Theory*, pp. 380-381.

24. Ibid., p. 393.

25. Percy W. Bridgman, in *A. Einstein*, ed. Schilpp, pp. 335-354, notes the fundamental theoretical impact on both relativity and quantum theory of the speed of light as well as the gravitational constant. Yet this invariance is rooted in a simple observational act.

26. Cassirer, *Einstein's Theory*, p. 374.

27. Ibid., p. 375.

28. *P.S.F.*, III, 436.

29. Ibid., p. 437.

30. Ibid., p. 438.

31. Cassirer, *Einstein's Theory*, p. 402.

32. Ibid., p. 401.

33. Ibid., p. 404.

34. E. Cassirer, *Substance and Function,* trans. W. C. and M. C. Swabey (New York: Dover, 1953). See also chap. 3, this study.

35. I. K. Stephens, "Cassirer's Doctrine of the *A Priori," Philosophy of Ernst Cassirer,* ed. Schilpp, pp. 156-158.

36. See T. S. Kuhn, *The Structure of Scientific Revolutions* (Chicago: Univ. of Chicago, 1962).

37. See P. Duhem, *The Aim and Structure of Physical Theory,* trans. Philip Weiner (Princeton, 1954); see also S. Toulmin, *Foresight and Understanding* (Bloomington: Indiana Univ. Press, 1961).

38. Cassirer, *Einstein's Theory,* p. 366.

39. *P.S.F.,* III, 446.

40. "Über die Entstehung und bisherige Entwicklung der Quantentheorie," Nobel Prize lecture, *Physikalische Rundblicke* (Leipzig, 1922), p. 148 ff., in ibid.

41. *P.S.F.,* III, 446-447.

42. A. Einstein, *Relativity: the Special and General Theory* (Chicago: Regnery, 1951), p. 37.

43. *P.S.F.,* III, 473.

44. Ibid., p. 472.

45. Cassirer, *Einstein's Theory,* p. 396.

46. Ibid., pp. 448-449.

47. E. Nagel, *The Structure of Science* (New York: Harcourt, 1961), p. 111.

48. See chap. 3, sec. 3, this study.

49. F. S. C. Northrop, "Introduction" to W. Heisenberg, *Physics and Philosophy* (New York, 1958), p. 7.

50. Ibid., p. 8.

51. P. Frank, *Philosophy of Science* (Englewood Cliffs, N. J.: Prentice-Hall, 1957), p. 220.

52. Ibid., p. 222.

53. H. Margenau, "Preface" to Cassirer, *Determinism and Indeterminism.*

54. Cassirer, *Determinism and Indeterminism,* p. 193.

55. Ibid.

56. Heisenberg, *Physics and Philosophy,* p. 186.

57. F. S. Northrop in ibid., p. 10.

58. A. Einstein, *Out of My Later Years* (New York: Philosophical Library, 1950), p. 91; see also Nagel, *The Structure of Science,* p. 310.

59. Cassirer, *Determinism and Indeterminism,* p. 188; also Frank, *Philosophy of Science.*

60. P. A. M. Dirac, *Principles of Quantum Mechanics* (Oxford, 1930), p. 23.

61. Nagel, *The Structure of Science,* p. 309.

62. Cassirer, *Determinism and Indeterminism,* p. 191.

63. Ibid., pp. 191-192.

64. Ibid., p. 189.

65. Ibid.

66. Ibid., p. 190.

67. Ibid., p. 191.

68. Ibid., p. 135; also see Nagel, *The Structure of Science*, p. 115.

69. Cassirer, *Determinism and Indeterminism*, p. 196; Nagel, *The Structure of Science*, pp. 106-110.

70. A. Einstein and L. Infeld, *The Evolution of Physics* (New York: Simon and Schuster, 1938), pp. 295-310.

71. Cassirer, *Determinism and Indeterminism*, p. 192.

72. DeBroglie states that quantum physics is at "that stage of a pure and simple cataloging of facts and establishing empirical laws," L. DeBroglie, *The Revolution in Physics* (New York; Noonday, 1953), p. 287.

73. Cassirer, *Determinism and Indeterminism*, p. 195.

74. Ibid.; Ernest Nagel states that the functional use of such terms within theories suggests that we ban the word "real" so that we never misconstrue the methodological utility of such a concept or confuse its status of theoretical invariance with substantial existence, Nagel, *The Structure of Science*, pp. 150-151.

75. N. Bohr, *Atomic Physics and Human Knowledge* (New York: John Wiley, 1958), pp. 59-60.

76. H. Margenau, "Preface" to Cassirer, *Determinism and Indeterminism*, p. xii.

77. Cassirer, *Determinism and Indeterminism*, p. 60.

78. F. S. C. Northrop, "Introduction" to W. Heisenberg, *Physics and Philosophy*, p. 17.

79. Ibid., p. 19.

80. E. Nagel, *Logic Without Metaphysics* (Glencoe, Ill.: Free Press, 1956), pp. 124-125.

81. E. Cassirer, *The Problem of Knowledge*, trans. W. H. Woglem and C. W. Hendel (New Haven: Yale, 1950), pp. 110-111.

82. Cassirer, *Determinism and Indeterminism*, p. 193.

83. Ibid., p. 191.

84. Ibid., p. 194.

85. Ibid.

86. *P.S.F.*, III, 479.

NOTES TO CHAPTER 5

1. M. Buber, *Between Man and Man* (New York, 1965), p. 119 ff.

2. See chap. 4 this study.

3. P. Frank, *Philosophy of Science* (Englewood Cliffs, N. J.: Prentice-Hall, 1957), pp. 339-341.

4. Ibid., p. 340.

5. Ibid., p. 352.

6. H. Weyl, as quoted by F. J. Dyson in *The World of Mathematics*, ed. J. Newman (New York, 1956), p. 1831.

7. E. Cassirer, *Einstein's Theory of Relativity*, trans. W. C. and M. C. Swabey (New York: Dover, 1953), p. 449.

8. Ibid., p. 450.

9. See S. Langer's development of this theme in respect to the arts, *Feeling and Form* (New York, 1953).

10. *Philosophy of Symbolic Forms* (New Haven: Yale, 1953-57) Vol. II: *Mythical Thought* (1955).

11. For an elaboration of the cultural implications inherent in the distinction between discursive and nondiscursive symbolism, see S. Itzkoff, *Cultural Pluralism and American Education* (Scranton: Intext, 1969).

12. *P.S.F.*, I, 190.

13. E. Cassirer, *An Essay on Man* (New Haven: Yale, 1962), p. 135.

14. Ibid., p. 136. Susanne Langer, following Karl Bühler and Philip Wegener, notes that the structure of language gradually manifests itself through a process of modifying and emending one-word sentences. But the existence of a language structure does not necessarily imply "generality." For this to develop, the circumstances of life must lead to the formation of logical analogues—"faded metaphors." These become the general abstractive words from which new discursive associations may grow. *Philosophy in a New Key* (Cambridge, Mass., 1957), pp. 135-143.

15. Otto Jespersen, *Language: Its Nature, Development and Origin* (London, 1922), p. 429.

16. *P.S.F.*, I, 199.

17. Ibid., p. 193.

18. See M. Buber, *Between Man and Man;* J. Piaget, *The Language and Thought of the Child* (Cleveland, 1955); E. G. Schachtel, *Metamorphosis* (New York, 1959).

19. *P.S.F.*, I, 267-268. See also N. H. Tur-Sinai for a confirmation of this point, "Language: An Enquiry into its Meaning and Function," in *The Origin of Language*, ed. R. A. Anshen (New York, 1957), pp. 41-79; also B. L. Whorf, *Language, Thought and Reality* (New York, 1956), pp. 242-243. The word means "He invites people to a feast."

20. *P.S.F.*, I, 280.

21. Ibid., p. 285.

22. Ibid., p. 292.

23. Ibid., pp. 291-292.
24. W. Wundt, *Völkerpsychologie*, 2d ed. (Leipzig, 1904), II, 15 ff., quoted in *P.S.F.*, I, 293.
25. *P.S.F.*, I, 294.
26. Ibid., pp. 295-296.
27. Ibid., pp. 296-297.
28. Ibid., p. 298.
29. Ibid., p. 300.
30. Ibid.
31. Ibid., pp. 301-302.
32. *Selected Writings of Edward Sapir*, ed. D. G. Mandelbaum (Berkeley, 1958), pp. 18-19; *P.S.F.*, I, 308-309; see also B. L. Whorf, *Language, Thought and Reality*.
33. A. L. Kroeber, *Anthropology* (New York, 1948), p. 229.
34. Ibid., p. 244; also John Carroll, *Language and Thought* (New York, 1964).
35. A recent confirmation of this relationship between a differentiated social structure and the development of abstract and general words and concepts is given by John L. Fischer, "Syntax and Social Structure: Truk and Ponape," in *Socio-Linguistics: Proceedings of the U. C. L. A. Socio-Linguistics Conference 1964*, ed. W. Bright (The Hague, 1966), p. 178.
36. *Selected Writings*, ed. Mandelbaum, p. 23.
37. Kroeber, *Anthropology*, p. 236.
38. Ibid., p. 287.
39. See S. Langer, *Philosophical Sketches* (Baltimore, 1962).
40. *P.S.F.*, I, 294.

NOTES TO CHAPTER 6

1. 2 vols. (Cambridge, 1926).
2. *The Philosophy of Symbolic Forms*, trans. Ralph Manheim (New Haven: Yale, 1953-57), III, 209.
3. Ibid., p. 217, note. Cassirer and Goldstein were, in addition, brothers-in-law.
4. Ibid., p. 216.
5. Hughlings Jackson worked between 1860 and 1890.
6. *P.S.F.*, III, 210.
7. Ibid., p. 213.
8. Hughlings Jackson, in H. Head, *Aphasia and Kindred Disorders of Speech*, I, 34 ff., quoted in *P.S.F.*, III, 212-213.

9. *P.S.F.*, III, 213.

10. Ibid., p. 211.

11. Ibid., p. 214.

12. Head, *Aphasia*, I, 211 ff., quoted in *P.S.F.*, III, 214-215.

13. K. Goldstein, "The Nature of Language," *Language: An Enquiry into Its Meaning and Function*, ed. R. A. Anshen (New York: Harper, 1957), pp. 18-19.

14. *P.S.F.*, III, 220.

15. Ibid., 221.

16. A. Gelb and K. Goldstein, "Psychologische Analysen," *Psychologische Forschung* VI (1925), 152 ff., quoted in *P.S.F.*, III, 225.

17. W. Stern, *Psychology of Early Childhood up to the Sixth Year of Age*, trans. Anna Barwell (New York, 1924); W. and C. Stern, *Die Kindersprache;* K. Bühler, *Die geistige Entwicklung des Kindes* (Jena, 1929); K. Goldstein and M. Scheerer, "Abstract and Concrete Behavior," in *Psychological Monographs*, LIII, no. 2 (Univ. of Illinois, 1941).

18. Gelb and Goldstein, "Psychologische Analysen," quoted in *P.S.F.*, III, 228.

19. D. Westerman, *Wörterbuch der Ewe-Sprache* (Berlin, 1903), in *P.S.F.*, III, 230.

20. *P.S.F.*, III, 230.

21. Ibid., p. 229.

22. Ibid., p. 232.

23. Ibid.

24. Ibid., p. 233.

25. Cassirer, *P.S.F.*, III, 262; cf. H. Liepmann, *Über Störungen des Handelns bei Gehirnkranken* (Berlin, 1905).

26. *P.S.F.*, III, 263.

27. Ibid., pp. 263-264.

28. Ibid., pp. 267, 269, 270.

29. Ibid., p. 265.

30. Ibid., p. 271.

31. Ibid., p. 273. One can also visualize the impossibility for the apractic of playing chess.

32. Ibid., pp. 273-274.

33. Ibid., p. 270.

34. Ibid., p. 182.

35. Ibid., p. 257 n.

36. Ibid., pp. 256-257.

37. Ibid., p. 257.

38. Ibid., p. 275. See also J. Carroll, *Language and Thought* (Englewood Cliffs, N. J.: Prentice-Hall, 1964).

39. *P.S.F.*, III, 275.

40. Goldstein and Scheerer, "Abstract and Concrete Behavior," p. 22. See also Kurt Goldstein, *Human Nature in the Light of Psychopathology* (1938), especially pp. 69-84. Note also the work of Penfield and Roberts, which goes a long way toward a kind of brain mapping with regard to correlating function to specific regions of the brain: W. Penfield and L. Roberts, *Speech and Brain Mechanisms* (Princeton, 1959).

41. Goldstein and Scheerer, "Abstract and Concrete Behavior."

42. Goldstein, *Human Nature*, pp. 59-60.

43. *P.S.F.*, III, 276.

44. Ibid., pp. 183-184; quotation from W. Shakespeare, *Hamlet*, IV, sc. 4, ll. 36-39.

45. E. Cassirer, *An Essay on Man* (New Haven: Yale, 1962), p. 24.

46. *P.S.F.*, III, 277.

47. See K. Goldstein's rather poetic discussion of this problem, inspired to a great extent by Cassirer, in *The Organism—A Holistic Approach to Biology* (New York, 1939), especially pp. 470-474.

48. See chap. 2, sec. 3, this study.

NOTES TO CHAPTER 7

1. H. Kuhn, "Review of *An Essay on Man*," *Journal of Philosophy* XLII, no. 18 (Aug. 30, 1945), 497-504.

2. J. von Uexküll, *Theoretische Biologie*, 2d ed. (Berlin, 1938); *Umwelt und Innenwelt der Tiere* (1909; 2d ed., Berlin, 1921).

3. E. Cassirer, *An Essay on Man* (New Haven: Yale, 1962), p. 32.

4. Ibid., p. 33.

5. Ibid., p. 24.

6. Ibid., p. 25.

7. Ibid., p. 31.

8. D. Bidney, "The Philosophical Anthropology of Ernst Cassirer and Its Significance in Relation to the History of Anthropological Thought," in *The Philosophy of Ernst Cassirer*, ed. P. A. Schilpp (New York: Tudor, 1958), pp. 465-544.

9. Cassirer, *An Essay on Man*, p. 27.

10. Kuhn, "Review," p. 500.

11. See articles by David Bidney, Fritz Kaufmann, John Herman Randall in *The Philosophy of Ernst Cassirer*, ed. Schilpp, pp. 151-181.

12. I. K. Stephens, "Cassirer's Doctrine of the *A Priori*," in ibid., pp. 175-176.

13. See R. Hofstadter, *Social Darwinism in American Thought* (New York, 1959).

14. E. B. Tylor, *Primitive Culture* (1871; London, 1929); L. H. Morgan, *Ancient Society* (1877; New York, 1963).

15. C. Darwin, *The Variation of Animals and Plants under Domestication* (London, 1868).

16. G. Mendel first published his researches into inheritance in 1866. He was unrecognized for thirty-four years. DeVries delineated the modern theory of gene mutations in 1901, and Thomas Hunt Morgan's important synthetic work with Drosophila dates from 1910 on.

17. I. Pavlov, *Lectures on Conditioned Reflex* (New York, 1928), in W. Dennis, *Readings in the History of Psychology* (New York, 1948), p. 430.

18. J. B. Watson, "Psychology as the Behaviorist Views It," *Psychological Review* XX (1913), 158-177; also J. B. Watson, *Behaviorism* (Chicago, 1958).

19. W. MacDougall, *An Introduction to Social Psychology* (Boston, 1908). In the twenties MacDougall became a bitter opponent of the behaviorists' mechanistic view of man.

20. C. L. Hull, *Principles of Behavior* (New York, 1943); also *A Behavior System* (New Haven, 1952).

21. B. F. Skinner, *The Behavior of Organisms* (New York, 1938); *Science and Human Behavior* (New York, 1953); *Walden Two* (New York, 1948).

22. K. P. Oakley, *Man, the Tool Maker* (Chicago, 1957); examples of this view can be found in M. F. Ashley-Montagu, *Anthropology and Human Nature* (Boston, 1957), p. 120 ff.; W. LaBarre, *The Human Animal* (Chicago, 1954), passim; S. Arieti, "Some Basic Problems Common to Anthropology and Modern Psychiatry," *American Anthropologist* LIX (1956), 26-39; "Human Beginnings," *Man, Culture and Society*, ed. H. L. Shapiro (New York, 1956), pp. 4-8.

23. G. S. Carter, "The Theory of Evolution and the Evolution of Man," in *Anthropology Today*, ed. A. L. Kroeber (Chicago, 1953), p. 341; also H. J. Muller, "The Guidance of Human Evolution," in *The Evolution of Man*, ed. S. Tax (Chicago, 1960), p. 424.

24. See also V. G. Childe, *Social Evolution* (New York, 1951); A. Keller, *Societal Evolution* (New York, 1931); *The Social Psychology of George Herbert Mead*, ed. A. Strauss (Chicago, 1956).

25. L. White, *The Science of Culture* (New York, 1949); also *Evolution and Culture* (Ann Arbor, Mich., 1960).

26. R. Ardrey, *African Genesis* (New York, 1961); also *The*

Territorial Imperative (New York, 1966); K. Lorenz, *On Aggression* (New York, 1966); L. D. Darlington, *The Facts of Life* (London, 1953); T. Dobzhansky, *Mankind Evolving* (New Haven, 1962); P. Teilhard de Chardin, *The Phenomenon of Man* (New York, 1959).

27. G. de Laguna, *Speech: Its Function and Development* (New Haven: Yale, 1927), pp. 41-42, 49.

28. H. Hoijier, "The Relation of Language to Culture," in *Anthropology Today*, ed. Kroeber, pp. 554-573; "Language and Writing," in *Man, Culture and Society*, ed. Shapiro, pp. 202-203; see also R. Brown, *Words and Things* (Glencoe, Ill., 1958).

29. The psychologist Lloyd Morgan, as early as 1894, promulgated his own version of Occam's Razor, stating that in behavior we ought not attribute to a higher function what could be attributed to a lower psychological function: cited by E. Hilgard in *The Evolution of Man*, ed. Tax, p. 270.

30. L. Bloomfield, *Language* (New York, 1933), p. 73.

31. C. Morris, *Signs, Language and Behavior* (New York, 1955), p. 198.

32. Ibid., p. 201.

33. See B. F. Skinner, *Verbal Behavior* (New York, 1957).

34. S. Chase, *The Tyranny of Words* (New York, 1938); C. K. Ogden and I. A. Richards, *The Meaning of Meaning* (New York, 1938); A. Korzybski, *Science and Sanity* (Lancaster, Pa., 1941).

35. As exemplified in L. Wittgenstein, *Philosophical Investigations* (New York, 1953).

36. E. Thorndike, *Animal Intelligence* (New York, 1898); *The Psychology of Learning*, I; and *Educational Psychology*, II (New York, 1913).

37. W. H. Kilpatrick, *Foundations of Method* (New York, 1932); "A Reconstructed Theory of the Educative Process," *Teachers College Record* XXXII, March, 1931 (Revised January, 1935); *Philosophy of Education* (New York, 1951).

38. G. Kennedy, in *Classic American Philosophers*, ed. M. Fisch (New York, 1951), p. 332.

39. J. Dewey, "The Reflex Arc Concept in Psychology," in *John Dewey—Philosophy, Psychology and Social Practice*, ed. J. Ratner (New York, 1965), pp. 252-266 (from *Psychological Review*, July, 1896, pp. 357-370).

40. See J. L. Childs, *American Pragmatism and Education* (New York, 1956).

41. Dewey's first important writings in education bring this out: *A Child and the Curriculum* (1902) and *School and Society* (1899; Chicago, 1956).

42. J. Dewey, *Human Nature and Conduct* (New York, 1922) and "The Influence of Darwin on Philosophy," (1909), in *Classic American Philosophers*, ed. Fisch, pp. 336-344.

43. J. Dewey, *Logic: The Theory of Inquiry* (New York, 1938).

44. J. Dewey, *The Public and Its Problems* (New York, 1927); *Individualism Old and New* (New York, 1930).

45. See H. Benjamin, *The Saber Tooth Curriculum* (New York, 1939).

46. See for example his two perhaps most important books, *Democracy and Education* (New York, 1916) and *Art As Experience* (New York, 1934).

47. E. Cassirer, *The Problem of Knowledge* (New Haven: Yale, 1950), Part IV, planned between 1932 and 1940 and written in twenty weeks in the latter year.

48. Susanne Langer in *The Philosophy of Ernst Cassirer*, ed. Schilpp, pp. 379-400.

49. P. Rieff, *Freud: The Mind of the Moralist* (New York, 1961).

50. E. Erikson, *Childhood and Society* (New York, 1950); H. Hartmann, "Ego Psychology and the Problem of Adaptation," in *Organization and Pathology of Thought*, ed. D. Rapaport (New York, 1951); H. Werner, *Comparative Psychology of Mental Development* (New York, 1940).

51. G. Allport, *Personality and Social Encounter* (Boston, 1960); E. Fromm, *The Sane Society* (New York, 1955); K. Horney, *New Ways in Psychoanalysis* (New York, 1939); E. G. Schachtel, *Metamorphosis* (New York, 1959); H. S. Sullivan, *The Interpersonal Theory of Psychiatry* (New York, 1953).

52. K. Bühler, *Die geistige Entwicklung des Kindes* (Jena, 1929); J. Piaget, *The Language and Thought of the Child* (Cleveland, 1955); W. Stern, *Psychology of Early Childhood up to the Sixth Year of Age* (New York, 1924); L. S. Vygotsky, *Thought and Language*, ed. and trans. E. Hanfmann and G. Vakai (Cambridge, Mass., 1962).

53. See N. Chomsky, "Review of B. F. Skinner's *Verbal Behavior*," *Language* XXXV (Jan.-Mar., 1959), 26-58.

54. N. Chomsky, *Current Issues in Linguistic Theory* (The Hague, 1964), p. 111, also *Syntactic Structures* (The Hague, 1957).

55. N. Chomsky, "Cartesian Linguistics," paper presented to the Philosophy and Science Seminar, Smith College, May 3, 1967; also *Cartesian Linguistics* (New York, 1966).

56. Chomsky, "Cartesian Linguistics."

57. Ibid.

NOTES TO CHAPTER 8

1. See T. Dobzhansky, *Genetics and the Origin of Species*, 3d ed. (New York, 1951); R. A. Fisher, *The Genetical Theory of Natural Selection*, 2d ed. (New York, 1958); E. B. Ford, *Mendelism and Evolution*, 6th ed. (London, 1957).

2. See K. and C. Hayes, "The Cultural Capacity of the Chimpanzee," *The Non Human Primates and Human Evolution* (Detroit, 1955), p. 110 ff.

3. W. E. LeGros Clark, *History of the Primates* (Chicago, 1957), p. 89 ff.

4. R. Carrington, *A Million Years of Man* (New York, 1963), p. 52.

5. L. S. B. Leakey's recent uncovering of *Sinjanthropus* may push this back to the Pliocene. Richard Carrington argues that this creature is an *Australopithecus* (East African variety), *A Million Years of Man*, p. 81. See also W. Howells, *Mankind in the Making* (New York, 1959), p. 122.

6. Clark, *History of the Primates*, p. 164; also G. Clarke and S. Piggott, *Prehistoric Societies* (New York, 1967). Some anthropologists place Neanderthal man and Cromagnon man, his successor, in the same species, *Homo erectus*. Clarke and Piggott distinguish between the more primitive and classic neanderthaloids and more advanced, variable sapient types.

7. A period of either one million years in this traditional view or 400 thousand to 600 thousand in the Emiliani dating: C. Emiliani, "Notes on Absolute Chronology of Human Evolution," *Science* CXXIII (1956), 924-926.

8. The Swanscombe skull from the second interglacial period is very similar to homo sapiens and quite old. E. A. Hooton, *Up from the Ape* (New York, 1946), p. 359 ff.; Carrington, *A Million Years of Man*, pp. 115-116; Howells, *Mankind in the Making*, pp. 216-223.

9. L. K. Frank, "Comments on Genetic Evolution," *Daedalus* XC (1961), 459. The Mt. Carmel group in Palestine may have been a Neanderthal-Sapiens cross: Hooton, *Up from the Ape*, pp. 333-338.

10. L. Eisley, *The Immense Journey* (New York, 1957), p. 90, first attributes this view to A. R. Wallace.

11. Carrington, *A Million Years of Man*, p. 56.

12. Emiliani, "Notes on Absolute Chronology"; also "Dating Human Evolution," in *The Evolution of Man*, ed. S. Tax (Chicago, 1960), pp. 57-66. The work of Dr. Emiliani is also discussed in Eisley, *The Immense Journey*, p. 113 ff. Dr. Rhodes Fairbridge of Columbia

University has hypothesized that the Pleistocene endured for 400 thousand years, *New York Times*, Sept. 8, 1959.

13. On the autonomy of the gene, see H. J. Muller, "The Darwinian and Modern Conception of Natural Selection," *Proceedings of the American Philosophical Society*, 1949, pp. 460-463.

14. G. G. Simpson, "Tempo and Mode in Evolution and the Meaning of Evolution," also "Rates of Evolution in Animals," in *Genetics, Paleontology and Evolution*, ed. G. Jepson (Princeton, 1949).

15. E. Mayr, "Comments on Genetic Evolution," *Daedalus*, XC (1961), 467. Mayr attributes the stability of man's large brain to natural selection for cooperativeness in small societies.

16. See Eisley, *The Immense Journey*, p. 107 ff.

17. Ibid., p. 79 ff.

18. See E. M. Thomas, *The Harmless People* (New York, 1959) Carleton Coon, *The Origin of Races* (New York, 1962), attributes no special morphological significance to Boskop man.

19. S. Wright, "Adaptation and Selection," in *Genetics, Paleontology and Evolution*, ed. Jepson, p. 381; also "Population Structure in Evolution," in *Proceedings of the American Philosophical Society*, 1949.

20. H. J. Muller, "Redintegration of the Symposium," *Genetics, Paleontology and Evolution*, ed. Jepson, pp. 433-434.

21. B. Rensch, *Evolution above the Species Level* (New York, 1959). Rensch examines a number of traditional evolutionary problems from the standpoint of neo-Darwinism. His books constitute an attempt to muster, through the traditional schematism, evidence supportive of "progressive evolutionary" trends.

22. R. Goldschmidt, *The Material Basis of Evolution* (New Haven, 1940).

23. J. C. Willis, *Age and Area* (Cambridge, 1922), also *The Course of Evolution* (Cambridge, 1940); Wright, "Adaptation and Selection," in *Genetics, Paleontology and Evolution*, p. 386. Orthodox views are represented by Rensch, *Evolution above the Species Level;* J. Huxley, *Evolution, the Modern Synthesis* (New York, 1943); T. Dobzhansky, *Genetics and the Origin of Species*, 3d ed.

24. Haeckel, *Gemeinverständliche Vorträge und Abhandlungen,* 2 vols. (Bonn, 1902); H. Driesch, *The Science and Philosophy of Organism* (London, 1908); P. Teilhard de Chardin, *The Phenomenon of Man* (New York, 1959); E. S. Russell, *The Directiveness of Organic Activities* (Cambridge, England, 1946); E. W. Sinnot, *Cell and Psyche* (Chapel Hill, N.C., 1950).

25. See G. S. Carter, *Animal Evolution* (London, 1954), pp. 50, 54 ff.; Huxley, *Evolution, the Modern Synthesis*, p. 99.

26. G. G. Simpson, *The Meaning of Evolution* (New Haven: Yale, 1967), pp. 149-150; J. Huxley, *Problems of Relative Growth* (London, 1932); Carter, *Animal Evolution*, pp. 212-214.

27. See Goldschmidt, *The Material Basis of Evolution*.

28. Simpson, *The Meaning of Evolution*, p. 236 ff.; also G. G. Simpson, *The Major Features of Evolution* (New York, 1953), pp. 190-193. In the latter, Simpson lists nine categories of preadaptation.

29. See Teilhard de Chardin, *The Phenomenon of Man*.

30. Simpson, *The Meaning of Evolution*, p. 149 ff.; *Tempo and Mode in Evolution* (New York, 1944), chap. 5.

31. G. B. S. Haldane, *The Causes of Evolution* (New York, 1932), p. 22 ff.; Fisher, *The Genetical Theory of Natural Selection*, p. 52 ff.

32. Simpson, *The Meaning of Evolution*, pp. 62-77.

33. Wright, "Adaptation and Selection," pp. 386-388, and Muller, "Redintegration of the Symposium," p. 430, both in *Genetics, Paleontology and Evolution*, ed. Jepson.

34. Sir G. deBeer, *Embryos and Ancestors*, 3d ed. (Oxford, 1958).

35. Ibid., pp. 63-64.

36. L. Bolk, *Das Problem des Menschenwerdung* (Jena, 1926); L. Eisley, *The Immense Journey*, pp. 107-126; M. F. Ashley-Montagu, "Time, Morphology, and Neotony in the Evolution of Man," in *Culture and the Evolution of Man*, ed. M. F. Ashley-Montagu (New York, 1962), pp. 324-342.

37. A. A. Abbie, "A New Approach to the Problem of Human Evolution," *Transactions, Royal Society South Australia*, vol. 75: 70-88. Abbie argues that these changes in morphology can be analyzed best by studying the embryological changes which occur in all hominoid forms from about seven weeks on. These changes give rise to differing juvenile forms and in turn develop their typical adult types, quoted in *Culture and Evolution of Man*, ed. Ashley-Montagu, pp. 332-335.

38. DeBeer, *Embryos and Ancestors*, p. 76.

39. Ibid., pp. 123-124; also see A. Keith, *A New Theory of Human Evolution* (London, 1949), pp. 192-201.

40. W. Howells, *Mankind in the Making*, pp. 216-223; also see Clarke and Piggott, *Prehistoric Societies*, p. 64 ff., on the sudden and as yet untraceable ancestry of homo sapiens.

41. A. T. W. Simeons, *Man's Presumptuous Brain* (New York, 1960). Simeons argues that though the cooperation of cortex and cerebellum with regard to motor behavior, e.g., learning to ride a

bike, can be transferred to the automatic functions of the brain (diencephalon, brain stem), the continued suppression by the cortex of the natural physical actions and responses consequent to the "messages" given by the diencephalon—the instincts—can eventually cause the build up of repression and the onset of psychosomatic disease; also see A. Koestler, *The Ghost in the Machine* (New York, 1968) for an explication of the Papez-MacLean theory of man's three brains: reptilian (automatic); lower mammalian (emotions of attack and flight); and sapient (symbolism and cognition). The problems of man's social life, e.g., war, crime, emotional illness, etc., are attributed to the incomplete integration of these morphological structures and their correlative behavioral imperatives.

42. S. Langer, *Philosophical Sketches* (Baltimore, 1962), pp. 21-23.

43. W. Elkin, "Social Behavior and the Evolution of Man's Mental Faculties," pp. 131-147, and L. Eisley, "Fossil Man and Human Evolution," pp. 300-323, in *Culture and the Evolution of Man*, ed. Ashley-Montagu.

44. Langer, *Philosophical Sketches*, p. 136.

45. See H. Blum, *Time's Arrow and Evolution* (Princeton, 1951).

46. See A. I. Oparin, *Origin of Life* (New York, 1953).

47. W. Elsasser, *The Physical Foundations of Biology* (New York, 1958), esp. chap. 4.

NOTES TO CHAPTER 9

1. E. Cassirer, *The Myth of the State* (New York: Doubleday, 1955).

2. See *Philosophy of Symbolic Forms*, trans. Ralph Manheim (New Haven: Yale, 1953-57), vol. II: *Mythical Thought*, 213 ff.; see also J. H. Randall, Jr., "Cassirer's Theory of History," in *The Philosophy of Ernst Cassirer*, ed. P. A. Schilpp (New York: Tudor, 1958), p. 691 ff. Randall criticizes Cassirer for his lack of economic and geographical concerns.

3. Susanne Langer's various writings are probably the only example of Cassirer's contemporary influence in philosophy.

4. Only in *P.S.F.*, III, do we see Cassirer approaching the problem of knowledge from a genetic and developmental standpoint, i.e., from the most primitive and inchoate forms of thought to the highest — mathematics and science. And only in *An Essay on Man* is the problem of knowledge examined from the standpoint of a general theory of culture based on a theory of human nature.

5. E. Cassirer, *The Problem of Knowledge,* trans. W. H. Woglem and C. W. Hendel (New Haven: Yale, 1950); *An Essay on Man* (New Haven: Yale, 1962).

6. B. Groethuysen, "Towards an Anthropological Philosophy," in *Philosophy and History* (The Ernst Cassirer Festschrift), ed. R. Klibansky and H. J. Paton (New York: Harper, 1963), pp. 77-89.

7. S. Langer, *Feeling and Form* (New York, 1953); *Philosophy in a New Key* (Cambridge, Mass., 1957).

8. Mrs. Langer later modified her terminology. She now refers to the "import" of a work of art, of its having "expressive form," somewhat like a symbol: *Problems of Art* (New York, 1957), p. 127.

9. W. G. Walter, *The Living Brain* (New York, 1963), chaps. 7, 8.

10. See E. Cassirer, *P.S.F.*, vol. II: *Mythical Thought;* also *Language and Myth* (New York, 1946).

11. C. Kluckhohn, *Culture and Behavior* (New York, 1962), pp. 286-300.

12. L. White, *The Science of Culture* (New York, 1949) and *The Evolution of Culture* (New York, 1959); M. D. Sahlins and E. R. Service, eds., *Evolution and Culture* (Ann Arbor, 1960), esp. p. 93 ff.

13. Cassirer, *An Essay on Man*, p. 228.

14. White, *The Science of Culture*, chap. 13; Sahlins and Service, *Evolution and Culture*, p. 93 ff.

15. As we have noted earlier, Susanne Langer is the preeminent exception.

16. B. Blanshard, as a consequence, sees Cassirer as a great scholar and researcher rather than a synthetic philosopher in the traditional sense: *Philosophical Review* LIV (1945), 509-510.

17. P. Frank, *Modern Science and Its Philosophy* (New York, 1955), p. 172 ff.

18. E. Nagel, *Logic Without Metaphysics* (Glencoe, Ill.: Free Press, 1956), p. 172.

19. The following have been most consistently identified as members: Rudolf Carnap, Herbert Feigl, Kurt Gödel, Hans Hahn, Otto Neurath, Karl Popper, Moritz Schlick, Friedrich Waismann.

20. Wittgenstein does state that since his own analysis raises the issue of the meaning of the correspondence of language and atomic facts and can only lead to metaphysical questions concerned with this mirroring, the propositions expressed in his *Tractatus* are nonsense.

21. *Allgemeine Erkenntnislehre* (Berlin, 1918).

22. Nagel, *Logic Without Metaphysics*, p. 220.

23. M. Schlick, *Space and Time in Contemporary Physics*, trans. H. L. Brose (New York: Oxford, 1920), pp. 85-86.

24. J. Passmore, *A Hundred Years of Philosophy* (New York, 1957), p. 378. As Passmore phrases it, Neurath plays Berkeley to Schlick's Locke.

25. Frank, *Modern Science and Its Philosophy*, p. 36.

26. Passmore, *A Hundred Years of Philosophy*, pp. 425-426.

27. (New York, 1953).

28. Passmore, *A Hundred Years of Philosophy*, p. 377; also Frank, *Modern Science and Its Philosophy*, pp. 31-34.

29. R. Von Mises, *Positivism* (Cambridge, Mass., 1951), pp. 91-98.

30. Passmore, *A Hundred Years of Philosophy*, pp. 379-382; Nagel, *Logic Without Metaphysics*, pp. 220-224.

31. Percy W. Bridgman's operationism was an attempt to do within physics what the earlier stage in logical positivism represented itself as doing for philosophy. This was to redesign the structure of physical theories in order to avoid all unobservable theoretical stipulations that could not in some manner be joined directly to experimental or observational activities. See his *Logic of Modern Physics* (New York, 1928) and *The Nature of Physical Theory* (Princeton, 1936). His program elicited significant criticisms. See Ernest Nagel, *The Structure of Science* (New York: Harcourt, 1961), pp. 45-46, 115, 270; also Carl G. Hempel, "The Theoretician's Dilemma: A Study in the Logic of Theory Construction," in *Philosophical Problems of Natural Science*, ed. Dudley Shapere (New York, 1965), pp. 31-52.

32. S. Barker, *The Philosophy of Mathematics* (New York, 1964), pp. 92-94.

33. Nagel, *Logic Without Metaphysics*, pp. 224-228.

34. Passmore, *A Hundred Years of Philosophy*, pp. 382-385.

35. R. Carnap, *The Logical Syntax of Language* (New York, 1934), p. 284; also see *Philosophical Problems of Natural Science*, ed. Shapere, pp. 8-11.

36. Frank, *Modern Science and Its Philosophy*, pp. 36-37.

37. Ibid., p. 85.

38. Ibid., p. 70.

39. R. Carnap, "Testability and Meaning" (1936-37), in *Readings in the Philosophy of Science*, ed. H. Feigl and M. Brodbeck (New York, 1953), pp. 69-70.

40. Nagel, *Logic Without Metaphysics*, pp. 228-231; also see R. Carnap, *Philosophy and Logical Syntax* (New York, 1935).

41. H. Feigl, "Unity of Science and Unitary Science" (1939), pp. 382-383, also G. Bergmann and K. W. Spence, "The Logic of Psychological Measurement," p. 104 ff., both in *Readings in the Philosophy of Science*, ed. Feigl and Brodbeck.

42. (Edinburgh, 1962).

43. E. Nagel and J. Newman, "Goedel's Proof," in *The World of Mathematics*, ed. J. Newman (New York, 1956), pp. 1668-1695.

44. Ibid., pp. 1669, 1685.

45. Barker, *The Philosophy of Mathematics*, p. 94 ff.

46. Nagel and Newman, "Goedel's Proof," p. 1685.

47. Ibid.

48. Ibid., p. 1694.

49. Ibid., p. 1695. See a critique of Nagel and Newman's conclusions by Hilary Putnam in *Philosophy of Science* XXVII (April 1960), 205-207; also Nagel and Newman's reply in *Philosophy of Science* XXVIII (April 1961), 209-211.

50. In his later years, Carnap turned away from the synthetic vision to a variety of concerns in the philosophy of science having to do with more discrete issues such as semantics, probability theory, symbolic logic, and the philosophy of physics.

51. (Cambridge, England, 1951).

52. E. Zilsel, "Physics and the Problem of Historico-Sociological Laws," in *Readings in the Philosophy of Science*, ed. Feigl and Brodbeck, p. 720.

53. *The Physical Foundations of Biology* (New York, 1958).

54. N. Bohr, *Atomic Physics and Human Knowledge* (New York, 1958), passim.

55. S. Toulmin, *Foresight and Understanding* (Bloomington, Indiana Univ. Press, 1961). Toulmin argues in the tradition of Duhem against views of scientific verification and truth which postulate the decisive results that are purportedly achieved when any one set of experiments or facts confirm or disconfirm a general theory. His view of the significance of science thus denies the validity of the so-called predictivist criterion of scientific truth, a position which implies objectivism and realism in knowledge. The highest function of scientific theories, rather than being to establish permanent truths, lies in uniting in theory disparate domains of experience. As can be seen, this position interprets the significance of science in terms similar to Cassirer's analysis of the universalizing trends of scientific thought. T. Kuhn, *The Structure of Scientific Revolutions* (Chicago, 1962).

56. See Charles Coulston Gillispie, *The Edge of Objectivity* (Princeton, 1960).

57. Kuhn, *The Structure of Scientific Revolutions*, pp. 42, 108.

58. Ibid., p. 83.

59. Ibid., pp. 106-108.

60. See Philip P. Wiener in *Philosophy of Science* XXV (1958), 298.

61. Kuhn, *The Structure of Scientific Revolutions*, p. 158.

62. W. Heisenberg, *Physics and Philosophy* (New York, 1958), pp. 163-166; H. Margenau, *Open Vistas* (New Haven, 1961), pp. 215-231.

63. Dudley Shapere wonders on this relativistic position how, given two equally adequate accountings of the facts, a choice can be made. The question itself assumes an objectivistic status for scientific theory. See *Philosophical Problems of Natural Science*, p. 27.

64. J. R. Oppenheimer, *Science and the Common Understanding* (New York, 1954).

Index

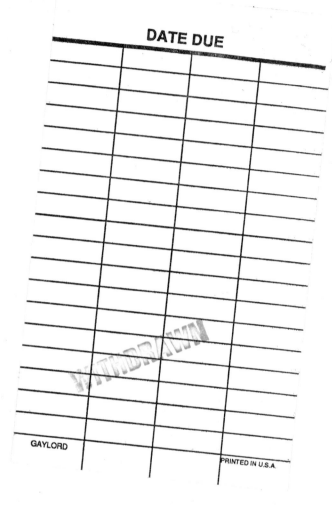